TSTI-Waco
3 6705 00025481 5
AF691860

PERGAMON INTERNATIONAL LIBRARY
of Science, Technology, Engineering and Social Studies
The 1000-volume original paperback library in aid of education, industrial training and the enjoyment of leisure
Publisher: Robert Maxwell, M.C.

YOUR SOLAR ENERGY HOME
Including Wind and Methane Applications

THE PERGAMON TEXTBOOK
INSPECTION COPY SERVICE

An inspection copy of any book published in the Pergamon International Library will gladly be sent to academic staff without obligation for their consideration for course adoption or recommendation. Copies may be retained for a period of 60 days from receipt and returned if not suitable. When a particular title is adopted or recommended for adoption for class use and the recommendation results in a sale of 12 or more copies, the inspection copy may be retained with our compliments. The Publishers will be pleased to receive suggestions for revised editions and new titles to be published in this important International Library.

Other Pergamon Titles of Interest

BELAIR et al
Aspects of Energy Conversion

BOER
Sharing the Sun

DUNN & REAY
Heat Pipes, 2nd Edition

KOVACH
Technology of Efficient Energy Utilization

MCVEIGH
Sun Power

MESSEL & BUTLER
Solar Energy

SCHLEGEL & BARNEA
Microbial Energy Conversion

SIMON
Energy Resources

SPORN
Energy in an Age of Limited Availability and Delimited Applicability

Related Journals Published by Pergamon

Energy
Energy Conversion
Solar Energy
Sunworld

TH
7414
.H68
1979

YOUR SOLAR ENERGY HOME

Including Wind and Methane Applications

by

DEREK HOWELL

(Past Chief Engineer,
Hawker Siddeley Automatic Handling Division)

PERGAMON PRESS

OXFORD · NEW YORK · TORONTO · SYDNEY · PARIS · FRANKFURT

Library
Texas State Technical Institute

U. K.	Pergamon Press Ltd., Headington Hill Hall, Oxford OX3 0BW, England
U. S. A.	Pergamon Press Inc., Maxwell House, Fairview Park, Elmsford, New York 10523, U.S.A.
CANADA	Pergamon of Canada Suite 104, 150 Consumers Road, Willowdale, Ontario M2J 1P9, Canada
AUSTRALIA	Pergamon Press (Aust.) Pty. Ltd., P.O. Box 544, Potts. Point, N.S.W. 2011 Australia
FRANCE	Pergamon Press SARL, 24 rue des Ecoles, 75240 Paris, Cedex 05, France
FEDERAL REPUBLIC OF GERMANY	Pergamon Press GmbH, 6242 Kronberg-Taunus, Pferdstrasse 1, Federal Republic of Germany

Copyright © 1979 D. Howell

All Rights Reserved. No part of this publication may be reproduced, stored in a retrieval system or transmitted in any form or by any means: electronic, electrostatic, magnetic tape, mechanical, photocopying, recording or otherwise, without permission in writing from the publishers

First edition 1979

British Library Cataloguing in Publication Data

Howell, Derek
Your solar energy home
1. Solar heating
I. Title
697'.78 TH7413 78-40824

ISBN 0 08 022685 X (Hardcover)
ISBN 0 08 022686 8 (Flexicover)

Printed and bound at William Clowes & Sons Limited Beccles and London

Contents

054869

CHAPTER 1

Solar Energy in the U.S.

Whether the U.S. finds itself as world leaders in Solar due to somebody crossing out the $2.50 price tag on a barrel of oil and writing $10, or, whether it was due to the relentless pressure from Solar enthusiasts, will never be known, but leaders they are.

Either way, the overnight price increase in '73 has dramatic ally altered the energy situation, and none more so than the Solar energy situation. Prior to '73, cheap oil easily outpriced Solar in all of the world except the most prolific areas of Solar radiation, and thus effectively undermined any substantial public interest. But, with the oil still increasing in price (taking all other energy costs with it) the 'interest' in Solar has leapt up. Today the interest comes not only from those in the more prolific Solar areas, but also from others in less blessed areas. 'Interest' being a mere euphemism for viability, and that a further euphemism for 'is it cheaper?', a question more fully answered in a later chapter, but one that can be generalized thus.

Fig. 1 shows the Northern Hemisphere with 3 bands of 'Sun hours per year', and, surprisingly in this subject of many surprises, the best band is *not* that nearest the equator, but the middle one! This band runs around the world between latitudes 15° and 35° covering the Americas' all the way up from Guatemala to Las Vegas. The second best is the equatorial band below Guatemala, and the third (lat. 35° to 55°) covers the remainder of the U.S. mainland.

Neglecting that the best band is not next to the equator, what has now happened in the U.S. is that rising oil prices have

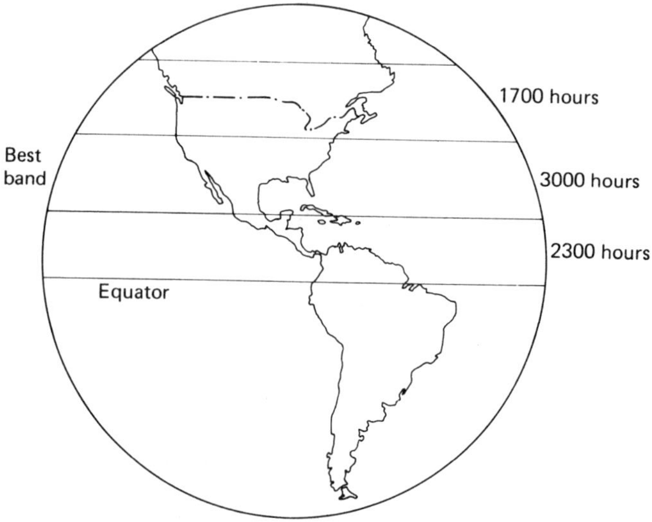

Fig. 1. The best solar band is not next to the equator, a peculiarity due to the tilt of the Earth relative to the Sun that gives longer hours of daylight to the 'best band'. But, since 1973 all three bands have become 'solar viable' bands.

moved the viability out from band 1 to include bands 2 and 3. There is even considerable activity in Montreal as they anticipate this sweeping viability arriving there. A comparison of the 3 bands is both interesting and encouraging: the best band averages 3000 hours of sunshine a year, the equatorial band 2300 and the U.S. mainland band 1700 hours of sunshine a year. Quite surprisingly, the best band has less than twice as much Sun as the mainland band. Now it should be apparent that if Solar was viable in the best band before the oil prices quadrupled, it must now be viable in the other 2 bands. Solar is viable over all of the U.S. and arrives 600 times faster than the present rate of consumption.

The days of cheap fuels are over, and, at last man is realizing that his temporary flirtation with the fossil fuels is ending. For our children, little more than a pleasant memory of a cheap, albeit polluting, fuel. But what is the alternative?

The great hope, atomic energy, has not mushroomed into prominence quite as expected, its contribution to our energy supplies is a long way overdue, and its development costs' (always ignored in calculating the cost of atom electricity), astronomically overspent. Yet, each day atomic energy becomes more and more suspect on safety and waste disposal grounds. Safety alone now emerges as a two-headed monster. Will there ever be an accident at an atomic station? Will somebody spirit away a few pounds of plutonium and blackmail a nation? And, unless the breeder reactor or atomic fusion is made to work, then the useable atomic fuels will run out at about the same time as the oil! There is considerable resistance building up in the U.S. monetary circles on this very point. Is it a good risk they ask, to finance atomic power stations that may not have sufficient fuels for their long-term operation?

The only other realistic alternative is Solar Energy: the world's only everlasting, non-polluting, ready distributed energy supply. (In this simple everyman's book there is a marked absence of mind-boggling numbers, but to qualify that 'everlasting' claim, the Sun is actually burning itself out. It is going at the rate of 4,000,000 tonnes (tons) a second and is expected to continue so for several thousand million years.) Quite a sobering thought as we feel its power arriving here on earth, but arrive it does, and it's free, but for the cost of collection. Nobody can turn it off, and nobody can up the prices, and it is here right on our walls and roofs waiting to be collected and used.

The technology that enables good use to be made of Solar is already well developed. Over the years that cheap oil, gas and coal kept Solar back, dedicated bands of researchers in many countries of the world experimented and perfected many aspects of the subject. For a change the technology is ahead of the demand; the reverse of the normal situation. And while there are many refinements to come, the last thing we need is for somebody to discover how to use the Sun.

As suredly though, we are doomed to a succession of such announcements in the ensuing years.

Along with other active centers in the world, U.S. Solar research can be traced in a continuous path from the end of World War II till today. Several Solar houses were built by Mass. Institute of Technology in the years '47, '49, and '59, with their proponents Prof. Hottel and Dr. Maria Telkes as active today as they were 30 years ago. And right in the middle of this academic effort, was a fine individual effort, when in '54 a Mr. H. R. Lefever built what was perhaps the first private Solar home in the U.S. Great activity also took place at the University of Wisconsin under the guidance of the world renowned Farrington Daniels. His text book, *The Direct use of the Sun's Energy* is a standard to this day. A Solar office building in Albuquerque, New Mexico, was operating in '56, more houses were built in Sante Fe in '58, Princeton in '59, and two more in Massachussetts in '60. All continue to add to the world store of practical Solar knowledge. Slightly overlapping these achievements, one Harry Thomason (much more about him later), started in '59 to build a wonderful series of six houses, all of which are operating to this day. In '63 Dr. Eric Farber at the University of Florida started an extensive program of research that continues strongly to this day. It now looks like he is coming up with some very good combined heating and cooling schemes.

To give a mind's eye view of the U.S. Solar scene through the years to present times, let us now look at some of the houses designed for Solar over the years and see briefly how they work. They are simple, understandable schemes – some of which have operated successfully for the last 15 or 20 years.

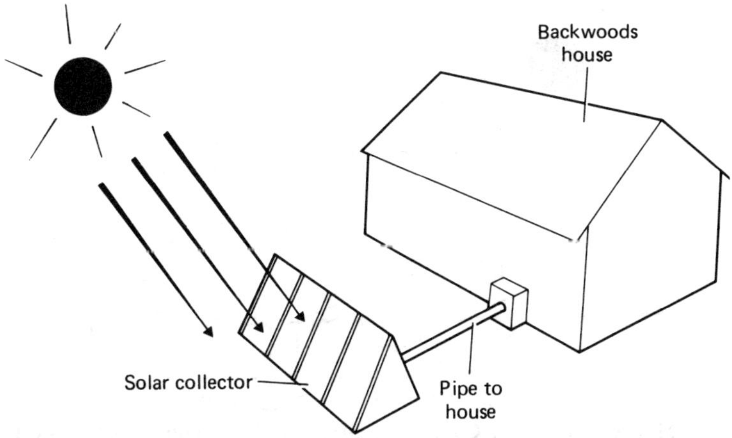

Fig. 2. Bliss Bungalow built at Amado, Arizona. Being out in the wilds, the appearance wasn't too important! Hot air is gathered behind glass in the long out-building and piped to heat a great bin full of small rocks. House air is circulated via the hot rocks to provide warmth. Electric heaters were installed but never used. This was the first 100% solar heated abode in the U.S., and the date? 1954!

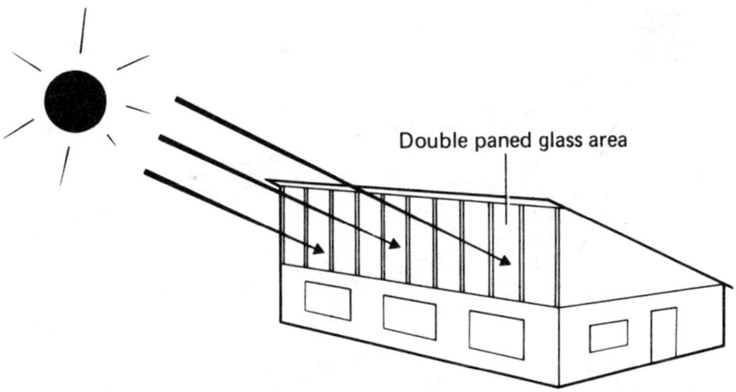

Fig. 3. Lefever House at Stoverstown, Penn. All the south-facing wall on the upper level is double paned, behind which a black plate collects the heat. Air is blown over the plate to take the heat into various heat store cupboards in the house. This is a 125 square meter house (1.325 square feet) that has been 50% solar heated since 1954.

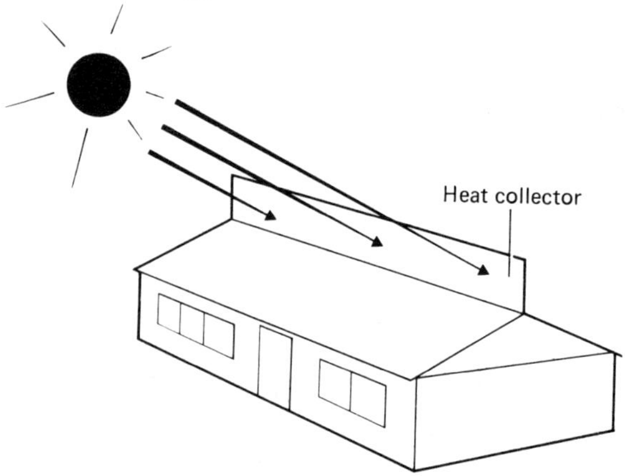

Fig. 4. Mathew House at Coos Bay, Oregon, 1967. This heat collector has water running through it from left to right. The resultant hot water being stored in a large tank under the living room so that the heat can be let out as needed, just where it is required. This 100% self-design, home-made system cuts the Mathews fuel costs by 85% according to the tests recently carried out by the Univ. of Oregon. Millwright Henry Mathew's approach seems to have been. 'I don't care if the equations say it can't work, I . . .!'

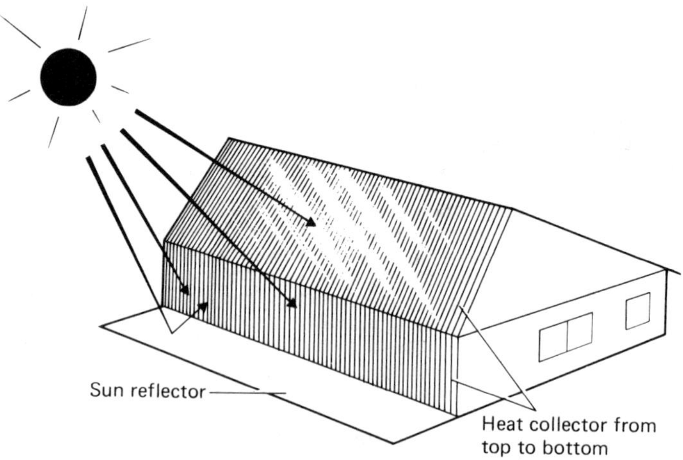

Fig. 5. Thomason Houses, Washington D.C. area 1960 onwards. This first simple home uses the same basic principle as the others of a trickling water collector. Water runs down a corrugated metal roof protected by a glass cover. Hot water collected at the bottom heats up a water and rock store for future use. These collectors often cover a complete roof; are quite cheap and get close to providing 100% heating, plus some cooling. Altogether an excellent cost effective system that we discuss in much detail later.

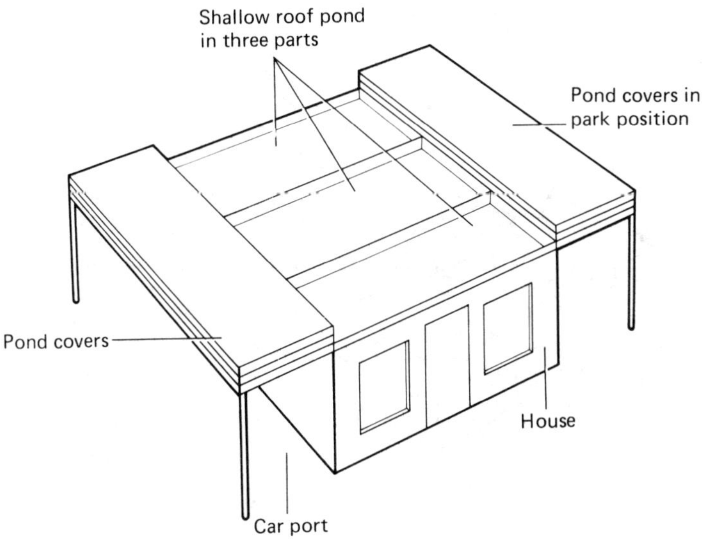

Fig. 6. Skytherm, or Hay House, Phoenix, Arizona. A shallow roof pond, exposed to the Sun, collects and stores the warmth. Insulating panels slide over the water to keep the heat in at night and thus direct the heat down into the house. The detailed operation (discussed later) also shows how the system is reversed in the summer to cool the house! A wonderful system that provides 100% solar heating and cooling.

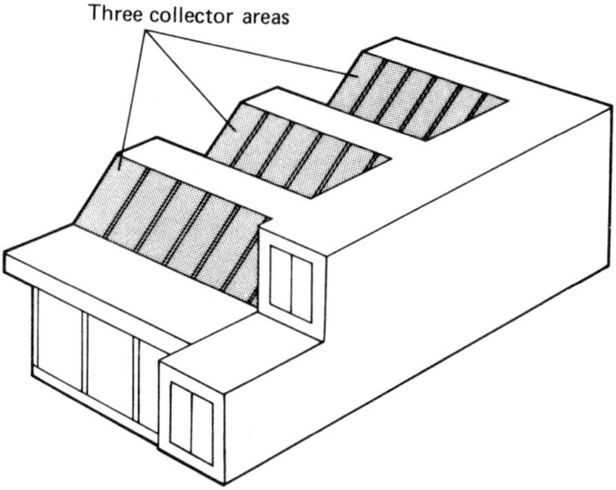

Fig. 7. Shoreline House at Westbrook, Conn. A straightforward use of collector panels that are readily available. Three areas of collection all piped into an underground water storage tank. Contribution ? Solar provides 45% of the heating and 75% of the domestic hot water, and that is in Connecticut!

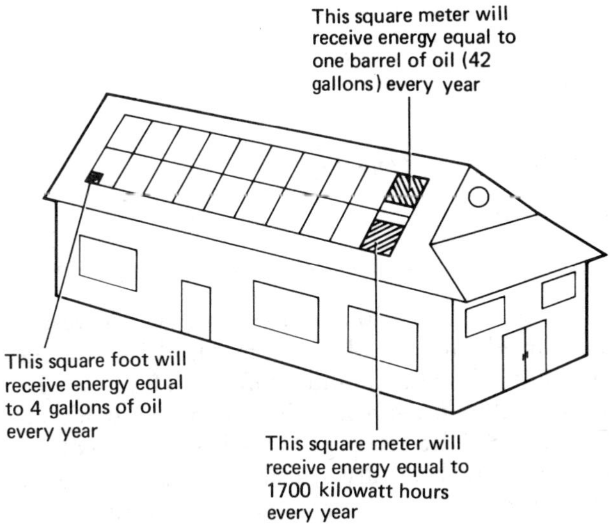

Fig. 8. The nearest most of us will get to drilling our own oil well! The house is in Illinois but the energy benefits could be yours!

CHAPTER 2

Solar Energy in the U.K. and Europe

The European contribution to the total World research effort has been considerable, and yet dominated by two names, Professor Thrombé in France and Professor Heywood in England. From about 1960 onwards, Felix Thrombé at his Laboratoire de L'Energie Solaire has produced some spectacular Solar furnaces (they are basically research tools for scientists). With his latest model, he concentrated 1 million watts of Solar Energy into a small enough area to melt and vapourize metal in seconds – all done by mirrors! Whilst this of no apparent use to ordinary Solar proponents, the work somehow led to the production of a series of Solar houses (built from 1967 onwards) centred around a very simple and elegant design concept evolved at the laboratory. Reference to, and operation of, the houses appear later on, on more than one occasion.

But even further back to 1947, the late Professor Harold Heywood started a series of experiments that now form the backbone of the U.K. contribution to the World store of Solar knowhow. Also, as a noteworthy educationalist, he inspired many student experimenters who now continue the work back in their native countries: a source of great joy to the Professor.

Professor Heywood, rightly dubbed the 'Father of U.K. Solar Energy', was fascinated by the inherent simplicity of the subject, and yet, sadly, was unable to see much of his pioneer work put to everyday use. A poignant scene was enacted recently when the wife and son of Professor Heywood attended the inaugural meeting of the U.K. Section of the International Solar Energy Society, a thriving and rapidly expanding group that is now the second largest in the World.

As a reminder of the work carried out in the U.K., France, Italy and other parts of Europe over a 25-year period of experimentation, we now look at some of the milestones along that road. Although some of the installations we show are nearly 20 years old, they are still in everyday use.

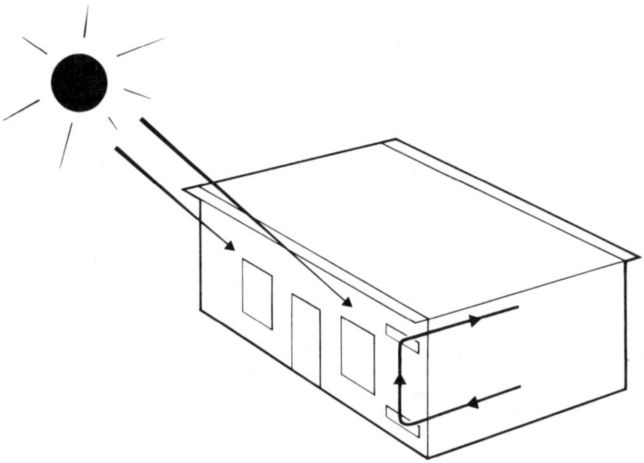

Fig. 9. House in Odeillo (Southern France). Built about 1967 as the first of a series based on the Thrombé/Michel principle of solar heating. The main front wall is massively thick and behind glass sheeting – the Sun stores heat in the wall which later gives up its heat to the house by an air flow up between the glass and the wall. Many variations and improvements have been made to succeeding designs, culminating in a grand three-storey version that Professor Thrombé built for himself in 1975. Like the restauranteur who eats in his own establishment, the Professor obviously has confidence in his own creation.

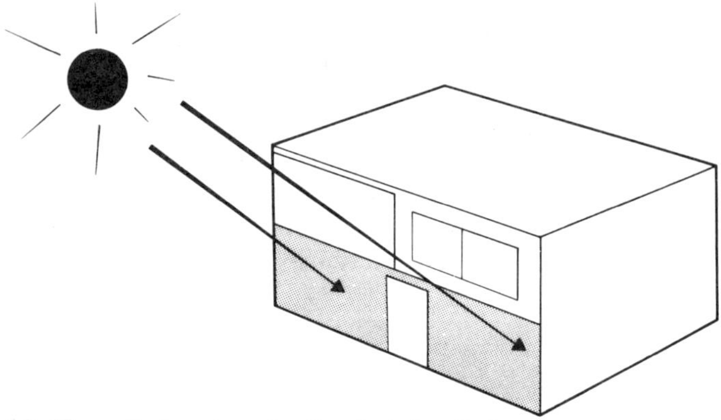

Fig. 10. House/Laboratory in Capri, Italy. Built as long ago as the late 1950's. The heat-collecting front wall is covered with flat radiator panels behind glass sheets. The water in them, after heating up, is stored in a large tank before being pumped round to heat the house in the usual manner.

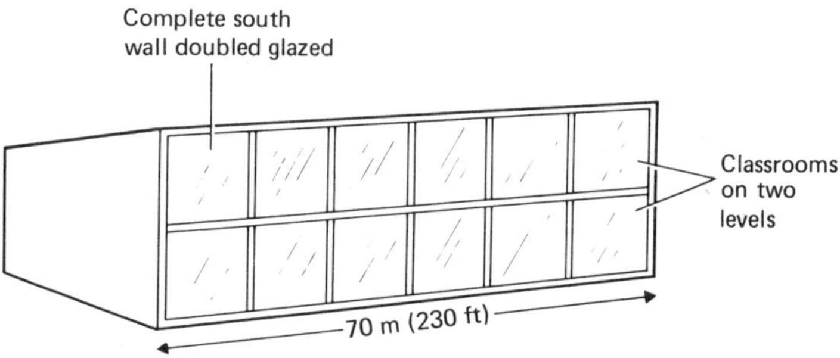

Fig. 11. A school in Wallesey, Cheshire. U.K. Designed in 1959 and perhaps the best known solar building in Europe. It has been kept within 3°C ever since without a recognizable heating system. All the heat comes from solar, plus waste heat from the lights and the warmth of the energetic little bodies of its occupants. When architect E. A. Morgan designed the building, higher authority insisted that he include a boiler. Imagine what a delicious day it was for Mr. Morgan, when after 4 years occupancy and not one day use from the boiler, he was allowed to remove it. The solar part works by allowing in maximum warmth via the south wall. As walls and floors are constructed of thick concrete they store the heat to use on sunless days. It is a sad reflection on British architects that the experiment has not been repeated.

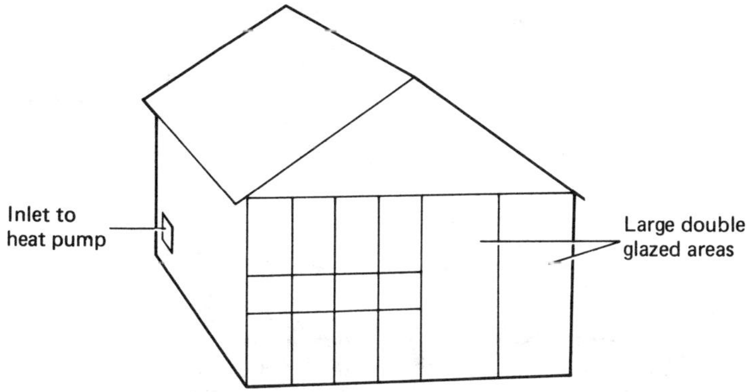

Fig. 12 Private house in Hertfordshire, U.K. Designed nearly 20 years ago by Architect Edward Curtis, and lived in ever since by himself and family. A good-looking contemporary design that gets its heating partially from large double glazed areas, and the remainder from a 'heat pump'. This is a device that *concentrates* a large volume of slight warmth into a small volume of greater warmth. This one has worked by actually getting its slight warmth from the outside air – which even in winter has some solar warmth in it.

[If you wish to know more about the house (and two more of Ed Curtis's solar designs), then send $2.00 for a 32-page book about them.

Send to:
North London Polytechnic (where Mr. Curtis is a principal lecturer)
Holloway Rd.,
London, N.7.]

Another famous heat pump is the one that heated the Royal Festival Hall for many years. It obtained its slight warmth from the Thames.

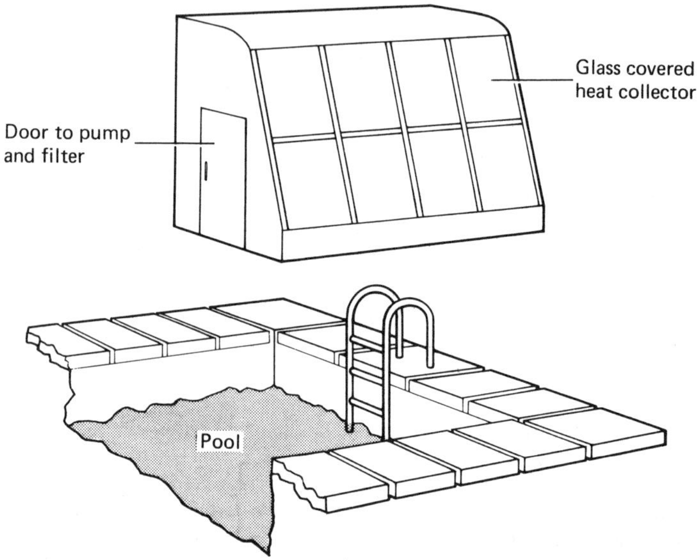

Fig. 13. Swim pool at East Grinstead, U.K. On the A22, south of Godstone, is 'Ye Olde Felbridge Hotel'. They have had a solar heated swimming pool since 1960, and it was constructed by self-confessed novices. It operates by circulating the pool water via the normal pump and filter arrangement, but, in addition, the water is taken up behind the glass and comes out warm for return to the pool. With luck Bill Shinn, one of the originators, will show you the works. The sight of a working system will encourage you to start your own system, especially when you realize that the same collector is suitable to heat your home.

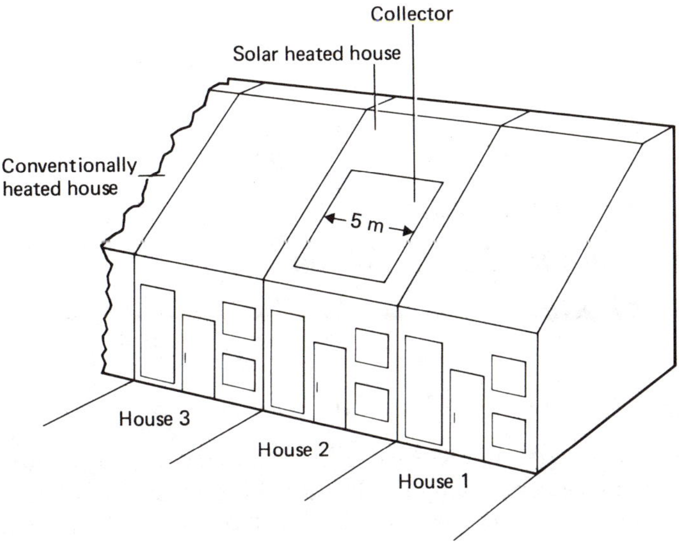

Fig. 14. Local Authority House at Milton Keynes, U.K. David Hodges – now in charge of the project at Central London Polytechnic – told us that one of the objectives was to convert a normal house. Another objective is to compare running costs with a similar house; they are using No. 4 in this block of 5 dwellings, and comparing it with No. 2. This is a large solar collector with water running inside it again. It is made from commercially available collector plates placed instead of tiles and glazed over with standard window fitments. A very large tank in the centre of the house stores sufficient how water to keep it all going for up to 2 days of continuous cloud cover; then, the standby gas boiler takes over.

CHAPTER 3

Solar Energy in Australia, New Zealand, South Africa and Canada

Australia (Lat. 11° to 39° South)

The vast majority of the Australian populace is wisely huddled close to the sea. By being so close to this enormous Solar store, they ensure that the main peaks and troughs of their weather are ironed out. In summer, the great mass of water can never rise in temperature to match that of the air with a consequent general cooling effect to the coastal areas. Likewise, in the winter the air temperature, in trying to drop well below that of the water, is to some extent held up by the relatively warm sea.

From this combination of being a country that is 95% in the 'best' Solar band, and having the main towns in the most climatically stable part of the country, the Australians enjoy one of the world's finest climates. And, as a direct result of these two factors, there is little need for continuous winter heating or summer cooling. But domestic hot water is required in abundance, and this is where Solar comes into its own in Australia – and as we have seen, based on very sound reasons.

Australia is one of the world's leaders in Solar water heating technology, having a history of Solar research going back at least to 1954; much of which has been led by the Federal funded Commonwealth Scientific, Industrial, Research Organization or CSIRO. Under the guidance of the world renowned Dr. Roger N. Morse, the laboratory has carved a name for itself second to none, and has been more than a little to do with the fact that the headquarters of the International Solar Energy Society (ISES) moved to Australia after having started in America.

The country's energy supply is based on coal and there are no nuclear stations in operation or envisaged. The target for Solar's contribution to this supply is 20% of the year 2000's heat requirements – which sounds more impressive when one considers that that is akin to the total 1975 consumption!

Since 1968 a small but efficient Solar industry has been built up based entirely on domestic water heating units, so that in excess of 20,000 units are now in operation in the country.

It is now Government policy that Solar is fitted to all houses and hostels in the Northern Territory. The next stage of Solar is aimed at providing the hot water that industry requires for its various processes. The collectors to achieve this are in the medium to large size, of which dozens are already in use in the country. These have collectors of from 25 to 200 sq. meters each (265 to 2120 sq. ft.) and are mostly on hotels, schools and Government establishments throughout the country.

Additional research is carried out in many Universities, and Queensland and New South Wales have installations that are used both as research and teaching vehicles.

With the rise in all fuel costs (including its own coal), interest is increasing, even in Australia, in Solar house heating and Solar cooling.

Australia is a country where Solar is steadily moving into the domestic and industrial situation. This is the result of a clear Federal policy extending back over many years, and one that was backed by reasonable funds. They started with domestic water heaters, are now moving into the production of Solar hot water for industry, and will soon add home heating and cooling. To give some idea of the progress to date, we have selected five typical installations.

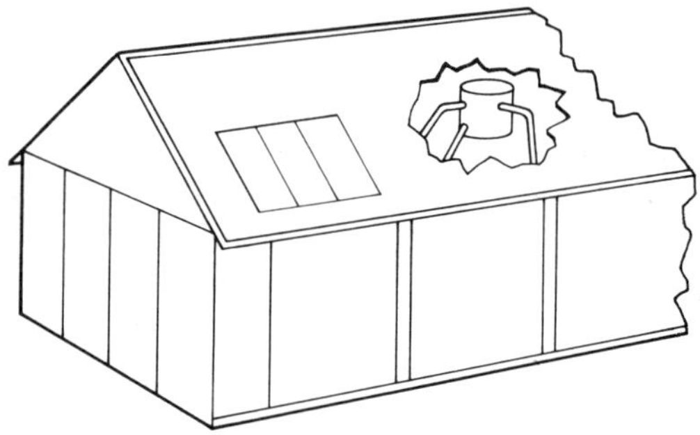

Fig. 15. Beasley Industries installation on a high pitch roof. They also have answers for other types of roofs.

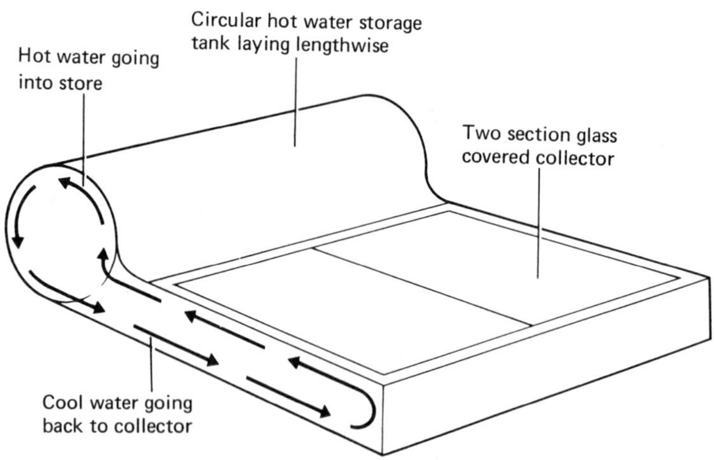

Fig. 16. This 'Sola-Ray' integrated collector and tank lays on the roof as one unit. It is a good example of many similar units in this very popular form.

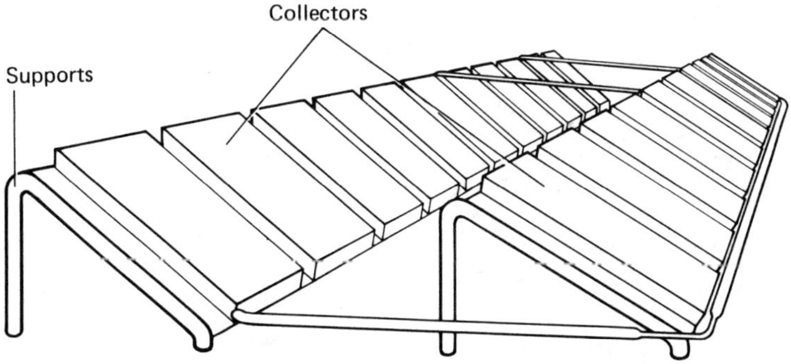

Fig. 17. Collectors as far as you can see. A 60 square meter (640 square feet) installation on the roof of the Sacred Heart College, Adelaide.

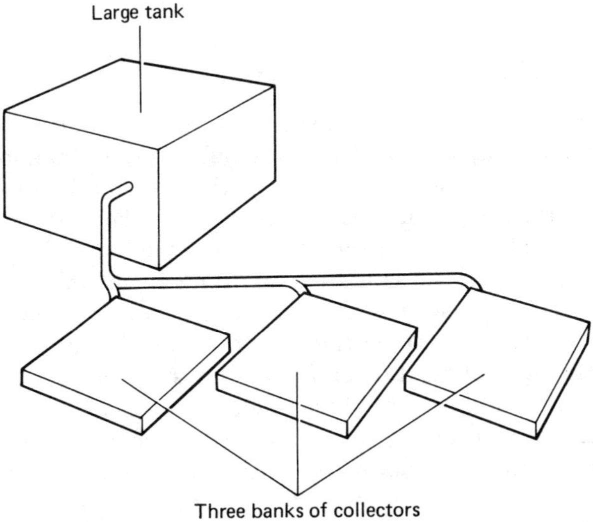

Fig. 18. The Ross Smith Hotel in Darwin with a 24 square meter (250 square feet) collector and tank on its roof. Note how the collectors are flatter to the roof for this location nearer the equator.

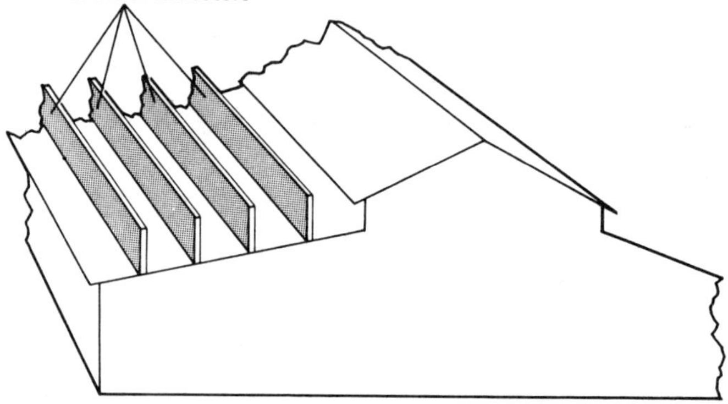

Fig. 19. This laboratory/office building of CSIRO now has a four-bank collector on its roof. They have no water in them – air circulates through and eventually heats the building.

New Zealand (Lat. 35° to 47° South)

New Zealanders live only just below the latitudes of the main population areas of Southern Australia, and as their towns are again near the coast, the climate they enjoy is similar. Predictably, they have adopted Solar in a similar way: to provide hot water. This is not a slavish following of one's larger neighbor's habits; it is following the situation's logical dictates.

The country also has its own Solar development program organized by the Department of Scientific and Industrial Research (DSIR) at Lower Hutt; and such is the nationwide interest, that nearly 15,000 leaflets have so far been sent out to Solar enthusiasts – that's a lot of interest in a country of less than 3 million inhabitants. This may have been because the DSIR carried out tests that showed that one would expect to obtain 50% to 60% of the year's hot water requirements from a single Solar installation in any part of the North or South Island.

With at least three local manufacturers supplying water heaters for home use, and many private units also in use, the number of satisfactory and profitable installations must run into many hundreds, if not thousands. Swimming pool heaters and Solar heated homes are also well known.

Being a country just in our Solar band 3, the very useful 2000 hours of sunshine a year gives good operation from single-glazed collectors, the same collector design being quite satisfactory all over the two islands.

An average water heater in New Zealand uses 3000 to 4000 units of electricity in a year in raising a 200-liter tank (53 U.S. Gallons) to about 50°C. (122°F). This amount of electricity can be reduced to half by the addition of a 4-square meter (43 sq. ft.) collector. Such a device will look just the same as those in the Australian section! Consequently, our following representations are of two Solar heated houses.

The one item that is shared with Australia is that they are both part of the same section of ISES, the combined Australia/New Zealand section.

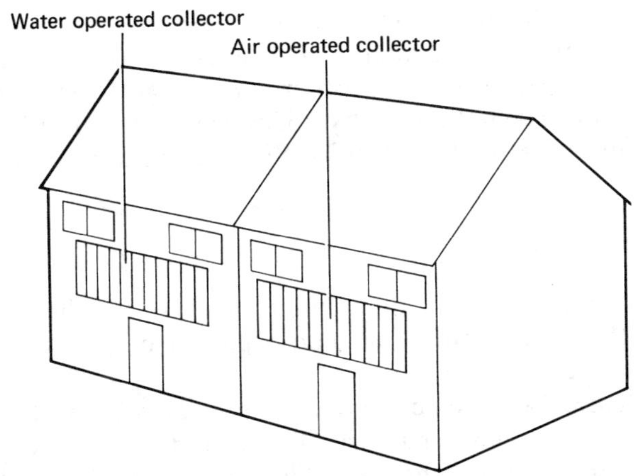

Fig. 20. These two New Zealand houses are similar looking, but one circulates water through the collectors, and the other air.

South Africa (Lat. 22° to 35° South)

With 100% of the country in our 'best' Solar band (15° to 35°), and the population again mainly near the coast, the same pattern for Solar emerges as in Australia: consequently, water heaters lead the parade. But the South African weather feature that produces rather cold winter nights, yet bright sunny days, is spurring the development of Solar heating for the home. This feature, which is particularly noticeable in Johannesburg, is a good example of considerable Solar strength without much *apparent* Solar strength; because there are other climatic influences at work.

As far back as the late 1950s, the South African Council for Science and Industry in Pretoria had a Solar development program; and, in a valuable and far-sighted paper presented to an International Conference in Rome '61, it outlined a method to standardize the measurement of Solar collectors. The very worthy objective of the paper was aimed at guiding this emerging world industry towards presenting all its claims to you, the users, in a like manner. Apples are difficult to compare with oranges, they contended!

More recently the general Solar effort has been rekindled by the energy crisis and many new research programs instigated. Not to be niggardly towards other country's efforts, one recent program has featured a tour by a U.S. expert to consult, lecture and confer at many of the leading South African centers.

Canada (Lat. 42° in E. and 40° in W. up to Arctic)

With the majority of the population close to the Southern border, they are virtually all encompassed by the top of our Solar band 3.

Canadian Solar research activity certainly goes back as far as 1950, when the University of Toronto did much work and study related to home heating. They grappled with such problems as: how does one keep the snow off the collector on the roof? To solve that they came up with one simple answer, just pump warm water round until it melts the under layer and it slides off! And another problem; how to store heat from the hot summer to the cold winter? Our first figure in this Canadian section shows a possible, and rather intriguing, answer.

More recently, the Canadian contribution to the world Solar scene has been blazoned by the Brace Institute at the McGill University in Quebec and its very lively Tom Lawand. While much of their work has been on Solar stills, to produce fresh water from salt, they also distribute detailed information of a very practical nature, all over the world. (The combination of the strong desert Sun and the lack of fresh water makes a very powerful case for the Solar operated still.) Their other activities include the running of a semi-permanent laboratory in

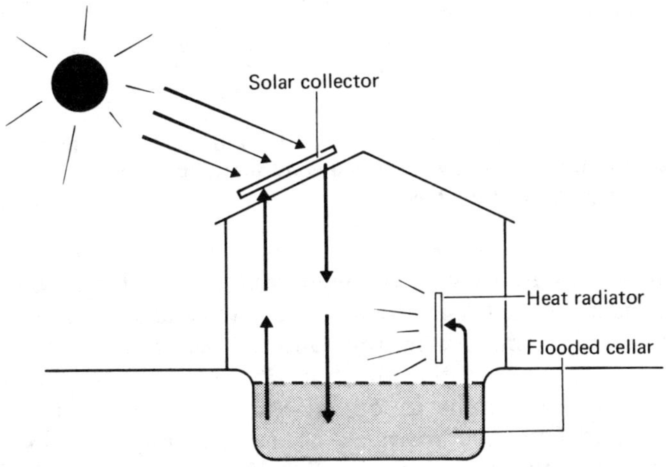

Fig. 21. Could your cellar be put to better use? If it was filled with water and used to store the summer warmth collected from the roof it can hold up to 6 months winter house heating. If you choose to use only half of the cellar, get expert advice on the strength of the dividing wall you put in, the water pressure is enormous. An inter season size store is no larger than a private swimming pool but so far home owners cannot quite comprehend that volume of water for anything than bathing.

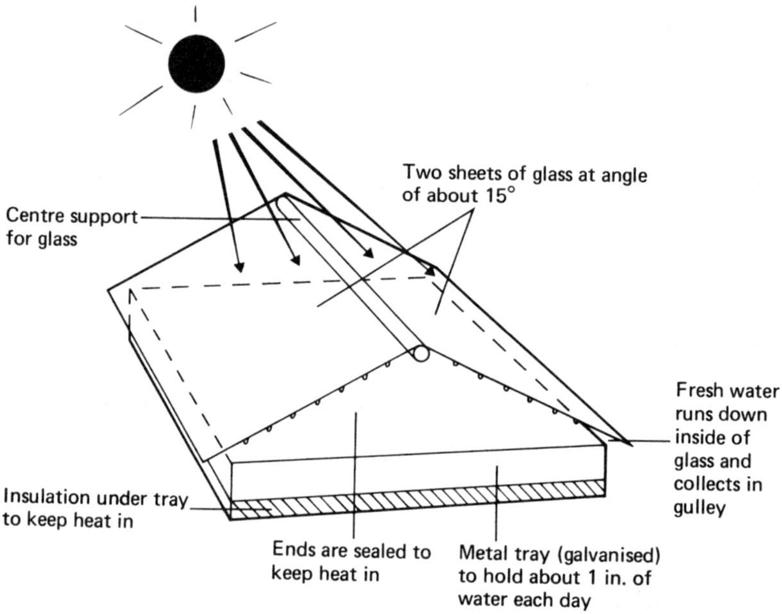

Fig. 22. This Brace Research Institute 'still' will convert 5 litres (1¼ gal.) of salt water into drinking water (on a good sunny day) for every 10 square meter (10 square feet) of glass.

Barbados, where they have Solar stills and a large windmill, to acting as consultants to many agencies around the world. Great activity (mainly in the Universities) ranges across the whole country, from the maritime Provinces in the East with several giant windmills in operation, to a dozen activities in Quebec and Ontario Provinces. Calgary, Winnipeg, Vancouver, and others in the West are all active in virtually every field and facet of Solar research.

In a recent change, the U.S. section of ISES renamed itself the 'American' section, in recognition of the long standing Canadian content and contribution. In addition, there is also an exclusive Canadian Solar Energy Society. This address, along with others, is included in the Directory.

The examples you have seen in the last two chapters give you a glimpse into the world of Solar Energy over the last 20 years. Each, in its own way, shows the end result of one particular train of thought. But, as many many other avenues were explored and recorded, it is in the laboratories and libraries of the world that the true wealth of Solar knowledge resides.

In these great storehouses are most of the answers that we require today. Solar is not a subject that requires inventing – it is a subject that requires *using*.

Our next chapter starts to show you how this inherently simple subject can be made to work for you.

CHAPTER 4

How to Collect Solar Energy

It is a certainty that there are many more forms of Solar Energy than most people ever imagined. That does not mean that there are several Suns, but that the energy appears in several different forms before it is recognized as Solar Energy. And it will be apparent as we progress through these chapters that, in fact, *all energy,* with the exception of atomic, is Solar Energy. A sweeping statement yet true to the extent that even coal, oil and gas are forms of Solar Energy.

In order to separate out the various forms of Solar Energy we will adopt three categories to work from:

1. Heat from the Sun's rays – that is, when there is little or no cloud cover.
2. Power from the Sun's light – any time except at night, cloudy or clear.
3. Power from air or water movement – any time day or night, cloudy or clear.

The mere setting out of these categories should in itself have underlined the opening statement. There are many sources of Solar Energy.

The first group – heat from the Sun's rays – is dominated by a multiplicity of methods designed to heat water. The water is then taken for domestic or central heating use. It is to the understanding of how best to collect the Sun's warmth that we now turn.

Many people actually have a crude form of Solar Collector in their possession but do not recognize it as such; they see it as an ordinary garden hose – for that is just what it is. But fill that hose with water, leave it out in the Sun for an hour and the water inside would become quite warm. The Sun's warmth has gone through the hose and collected in the water. Crude, but effective.

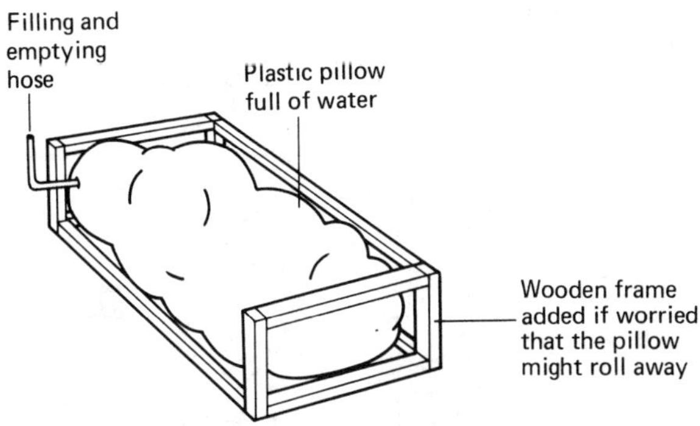

Fig. 23. Japanese 'Plastic Pillow' solar water heater – about 1,000,000 in use.

The above principle is used in a commercial water heater called the 'Japanese plastic pillow' – see Fig. 23. The alteration from the hose heater is that the collector instead of being in a long length is much shorter and shaped just like a pillow. But its operation is the same. The user fills the pillow with cold water in the morning, leaves it out in the sun all day, and draws off the hot water in the evening, generally to fill the traditional communal bath. While this appears to be a somewhat rudimentary system it is, nevertheless, acceptable to the Japanese, and they have about 1,000,000 of these 'pillows' in use. By no means an elegant arrangement, but certainly one of great simplicity.

A group of heaters similar to the 'pillows' are those that have the water in a more rigid structure. Refer to Fig. 24: this time the water is in a shallow metal tray painted black inside and fitted with a glass lid. These 'box heaters' are popular in rural

Russia and introduce us to two important principles.

First, the black base. After the Sun's rays have passed through the water, they hit the base, a color that increases the amount of heat that is absorbed. Black is a good absorber and white is bad, a fact that many people take advantage of by wearing white clothes in hot weather. This use of black is a crucial phenomena in Solar technology.

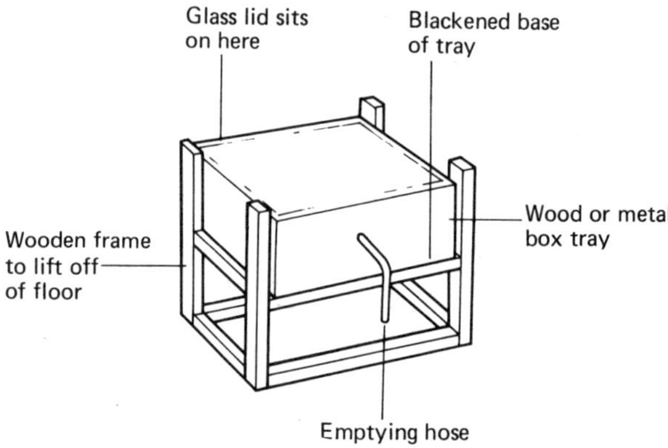

Fig. 24. 'Tray or box' type of solar water heater – popular in rural Russia.

Second, the glass lid. In simple terms it seems an entirely reasonable suggestion just to keep the heat in; but in essence we are seeing another phenomenon that is also crucial to the efficient collection of Solar Energy. One of the greatest benefactors from this effect are the horticulturists with their glass-houses or greenhouses, and so strong is the link that it is always referred to as the 'greenhouse effect'.

This is how it works. As the glass sheet is placed over the tray, a trap is created for the Sun's rays. A trap in the true sense of the word, just as a lobster pot is a trap for a lobster. Once the rays are in they cannot get out, an effect that is not directly to do with the glass (it would be the same if you turned the glass over). It has to do with the nature of the Sun's rays *after they have passed* through the glass. Once they are inside the box they hit the water, and in effect slow down – a perfectly reasonable occurrence – but, in so doing, they change their form (to what is called a longer wavelength) so that they can no longer pass out through the glass. Trapped. It is somewhat reminiscent of the monkey trap, where the victim is encouraged to put its hand inside a narrow-necked jar, and upon grasping the nuts finds that the *changed* forms of the hand traps it and prevents its escape.

These twin principles are so important that they form an integral part of virtually every Solar water heater. On the one hand, the blackened surface to absorb maximum heat, and, on the other, the glass lid to trap the maximum heat – simple yet vital functions.

You will undoubtedly be pleased to learn that there are more convenient collectors than the previous three, and our fourth is such a device. The collector itself is the most widely used in the world, and, due to its extensive development over many years, offers a multitude of variations. Fig. 25, shows a cross section; it would be about 2 meters (6½ ft.) high and 1 meter (3 ft.) wide, and be angled so that its full face (2m. by 1m.) pointed towards the sun.

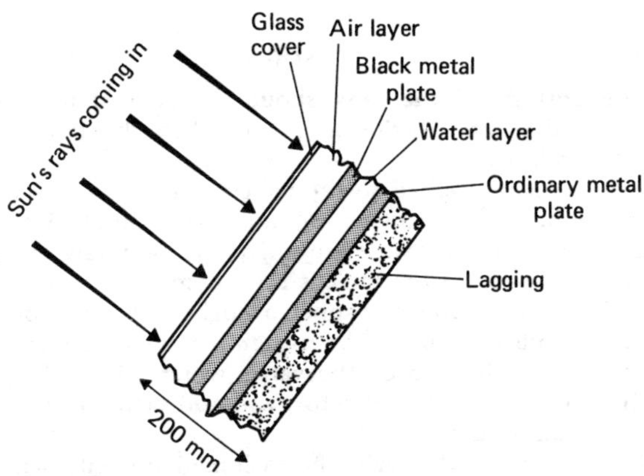

Fig. 25. 'Flat plate collector'. This 200 mm. (8 in.) thick cross section shows the sandwich construction.

The heat penetration through the 200mm. (8 inch) thick sandwich is:

1. Sun-rays go through the glass cover and the air layer to warm up the black metal plate.
2. The water layer is warmed by the black metal plate.
3. The ordinary metal plate (required to retain the water) is also, unfortunately, warmed up.
4. The heat insulation lagging keeps most of the heat inside the sandwich.

Thus, with the warmth safely in the water, it now has to be moved to where good use can be made of it. The simplest method of achieving this water movement is shown in Fig. 26, the 'thermosyphon' system. Its operation is based on one simple fact: hot water will rise to settle above a quantity of cooler water. As the collector heats up, the water in it rises out at the upper pipe and pushes its way into the top of the tank. This hot water then displaces some of the cold in the tank, pushing it down and out of the bottom. This heat-induced circulation is completed as the water, being pushed down the pipe, comes round the bottom and back into the collector.

Notes

1. The bottom of the tank should be at least 0.5 meter (20 ins.) above the top of the collector. Then the tank can eventually *fill* with hot water.
2. Just as 'heat' applied to the collector makes the water flow round in a clockwise direction, so may 'cold' applied to the collector make an anti-clockwide flow. By fitting a valve at X in Fig. 26, the hot water in the tank is prevented from reverse flowing. This is particularly liable to happen at night and loses stored heat.
3. As a modification to the system, a small electric pump can be fitted at X, instead of the valve, which allows the tank to be set above or below the collector, and still maintain the water flow in the required direction.
4. The pump should really be switched on only when the heat is ready to be collected, and the perfect guardian to achieve this is a sensing switch that gives a 'go' signal to the pump only when there is hot water in the collector.

5. A further refinement can be made by having an 'indirect' water system. Here the water going round and round the collector is always the same few liters, passing its warmth on to a second water circuit which contains the household water.
6. A new variation of the simple thermosyphon system uses 3 collectors (could be 2), each separately connected with 2 pipes each to the same tall tank. The idea takes note of the warm, medium, hot levels in a tank and uses a separate collector for each of matching

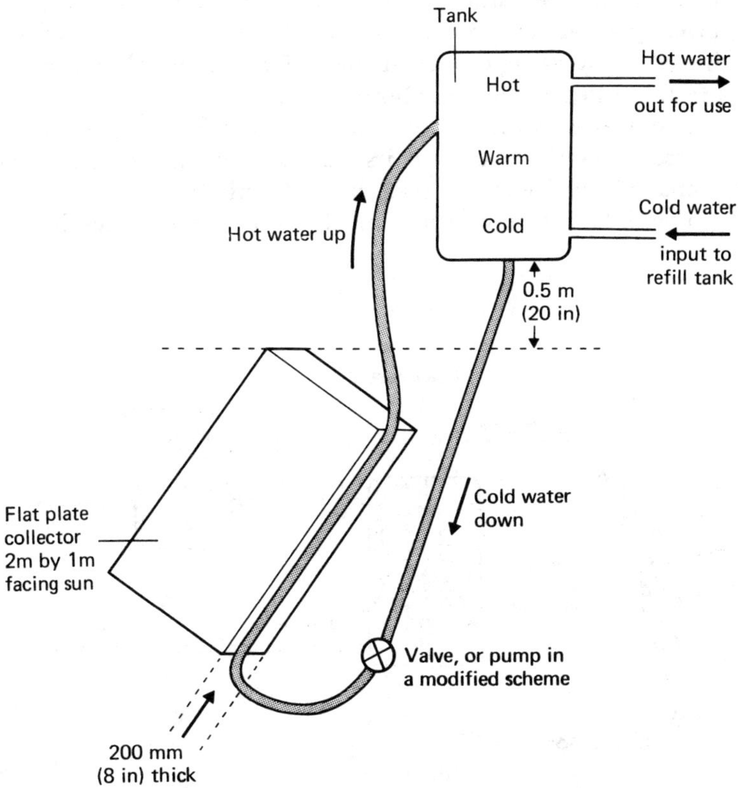

Fig. 26. The thermosyphon system of getting hot water out of the collector and into the house for use. This is a 'direct' method of heating water.

performance; the cheapest and crudest collector being used for the bottom or warm feed, up to the super version for the hot at the top of the tank. The tank is not divided into compartments.

This chapter is only concerned with the *principles* of collecting Solar Energy warmth into water. By putting together all that has so far been discussed, we arrive at the system shown in Fig. 27, a system that is not only practical but also in everyday use. Again, the variations abound – and some will be seen in the chapter about practical detailed systems – but we are looking at a system that most of the others can be related to.

Note the following points about Fig. 27.

1. The water going round the collector is a 'closed loop' arrangement, giving up its heat via the finned heat exchanger device in the Solar heated tank.
2. The Solar heating part – to the left of the dotted line –

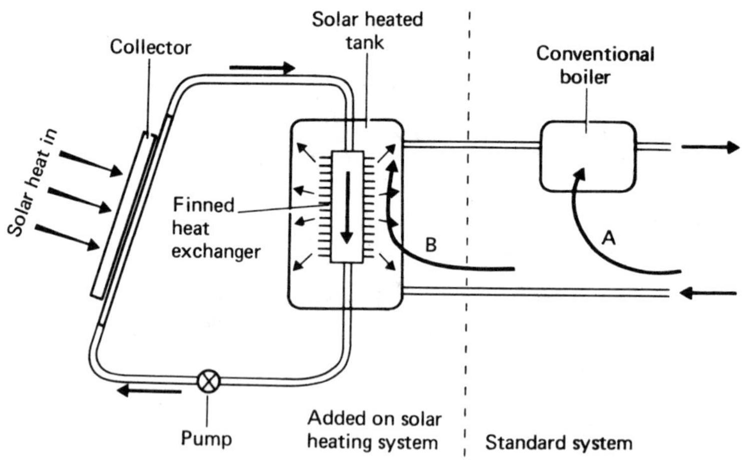

Fig. 27. Practical solar heating system. It often acts as a 'pre-heat' input to a conventional boiler. This is an 'indirect' method of heating water.

is best seen as an addition to a conventional heating and hot water system. Instead of cool water going back into a conventional boiler (via arrow A), it goes via the solar heated tank (via arrow B). At best the boiler need not switch on at all, but, more likely, it will only have to complete the heating operation. It is important to realize that it costs just as much to heat a quantity of water from cold to warm as it does to heat it from warm to hot; the last 20°C (68°F) rise costs the same as the first 20°C rise. In this context, the Solar Energy part can be seen as a valuable pre-heat device. That it does not provide all the energy input in no way detracts from its usefulness.

Thus far, we have dealt with schemes that gather heat into flowing water. But there are two more very worthy schemes that use fundamentally different techniques. Both schemes are combined collectors and stores, but as the store is the major part of the arrangement, they form a later chapter. Do not forget them when considering your choice of collector. One is the well-known system designed by Harold Hay in California, U.S., using shallow water ponds on the roof. The other is an equally famous scheme designed by Thrombé & Michel in France, using thick concrete storage walls and air flow. Both are elegant, simple arrangements.

Our second category of collection is: power from the Sun's light.

Whilst we all realize that when we switch on a light, we use power to supply that light, we do not recognize that power is being supplied to give us daylight. A light and power source that arrives even when the Sun is behind cloud and which illuminates with an intensity that makes a 100-watt bulb puny by comparison. This is a vast supply of Solar Energy that is quite separate from the heat energy; and this light energy arrives 12 hours a day 365 days a year!

For our purpose, we will split this light category into three groups, and while there are certain overlaps, they serve our purpose well as shown below.

(a) Photoelectric – produces electricity direct from the Sun's light.
(b) Photochemical – produces electricity, or, liquid and gaseous fuels by means of *non-living* chemical processes.
(c) Photobiological – produces food (animal/human fuel) and gaseous fuels by means of living organisms or plants.

The last two groups also share the term 'photosynthesis'. This is literally the building (synthesizing) by light. In the former group, the chemical reactions do the building; and in the latter, the plants do the building.

By starting with photoelectricity, we touch on the most publicized use of Solar Energy light – perhaps even of all Solar Energy. Every spacecraft uses it in the form of Solar cells to collect electricity for their various internal systems. The internal operation of these cells is a feat of electronic wizardry way beyond the stage we need to go to, but their external operation is simplicity itself. Show them the light and they produce electricity – evangelical electrics? The perfect collector – but at a price. It is sufficient to say that they are not likely to have a place in the home for some years to come. An interesting experimental use (cost no object) puts the Solar cells on the surface facing the sun of a normal hot water collector. The cells produce 15% of the energy as electricity and the remaining 85% heats the water.

The photochemical use of Solar Energy is, perhaps, the least well developed of the three groups of light users. This is unfortunate as it has an inherent characteristic with a special appeal – its ability to *store* Solar Energy for indefinite periods – a very valuable criterion that we shall elaborate upon later. There is hope that recent advances in chemical technology will soon filter through to this field and yield improvements. Already certain chemically filled cells are working that produce electricity, and others that produce a store of chemical fuels, but all have very low efficiencies. The actual collection of the Solar Energy into the processes is as simple as before – just show it the light.

The photobiological use of light by the green plants is the

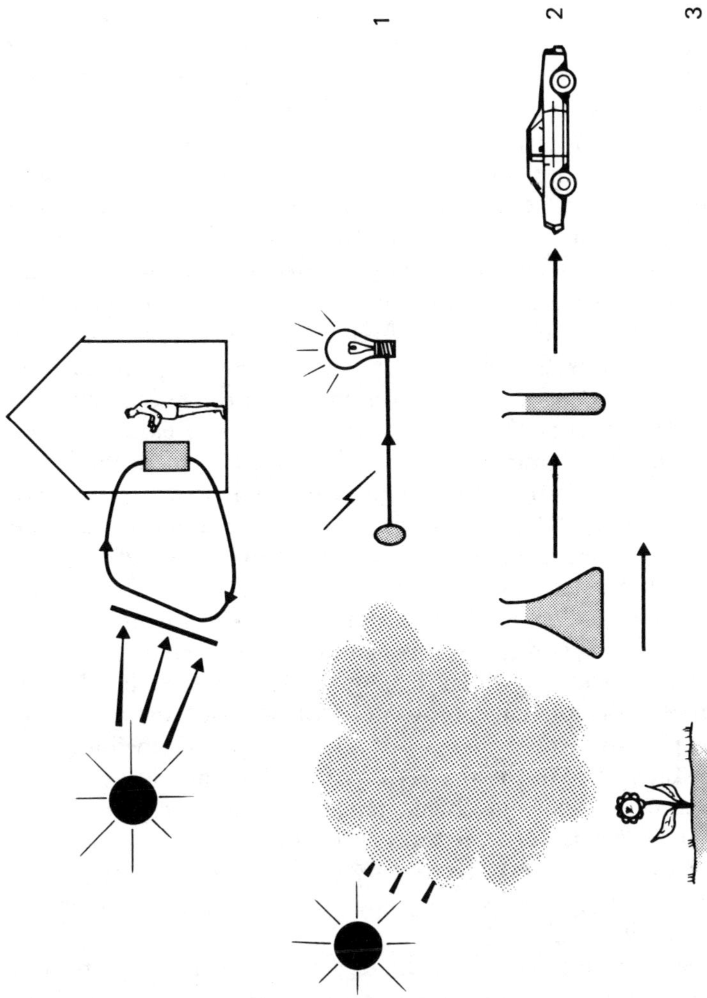

Fig. 28. The three basic ways of collecting solar energy when the Sun is in—from the light part of the rays.

greatest use of Solar Energy light in the world. Their leafy collectors take in vast quantities of Sunlight and use it to produce *energy rich* compounds from *energy low* ingredients. This immense collection and storage undertaking has continued since the start of life on earth, and is one that we already draw upon as we consume virtually any fuel or food.

This photobiological process is the fundamental life cycle of which plants are the center. At the start of the cycle, the basic elements, carbon, oxygen and hydrogen (as water and carbon dioxide) are built up into living things with the aid of Solar Energy. This is the magic of photosynthesis that stores Solar Energy in the vegetation. In the return part of the cycle, the consumption of the food and fuel takes the energy from the store and releases the elements back to their original state. It is to the more efficient utilization, interception and production of the stored energy that the researchers in the field are attracted. Already, methane gas is being produced from rubbish and vegetable waste; and, other fuels are coming along that we shall hear about later. To collect the Solar Energy into the plants and processes, once again, requires little or no effort; for, as you know, plants will even bend to seek out their Solar Energy. And as an act of final defiance they collect at maximum efficiency when the Sun is 'watery weak'.

Our third group of Solar Energy collectors are the devices that harness the physical movement of air and water and use them to produce energy. It is not always appreciated that wind is Solar Energy. Wind is the result of cool air moving in to replace warm air that has been caused to rise away from the ground by the action of the Sun. If it is any consolation to you, the next time you are buffetted by it, the wind you experience in your area is generally the result of somebody else's warm spell. It's an ill wind . . .

All water movement over land in the form of rivers and mountain stream is also Solar Energy; the water is merely returning from where it came – the sea. But, it is the Sun in the first place that keeps this other life cycle going. By evaporating the surface of the sea, the Sun draws the water up as clouds,

Fig. 29. The way to collect solar energy at night! A wind or watermill can operate 24 hours a day.

which then drift over to the high land and deposit themselves as rain. This rain gathers in gulleys and eventually forms rivers as it runs back to the sea. The lifting of water is the greatest use of Solar heat there is.

In both cases – air and water – it is the interception of these forces as they attempt to return to equilibrium that provides a tapping off point for yet two more forms of Solar Energy – in the former, the air rushing in to fill a void, and in the later, the

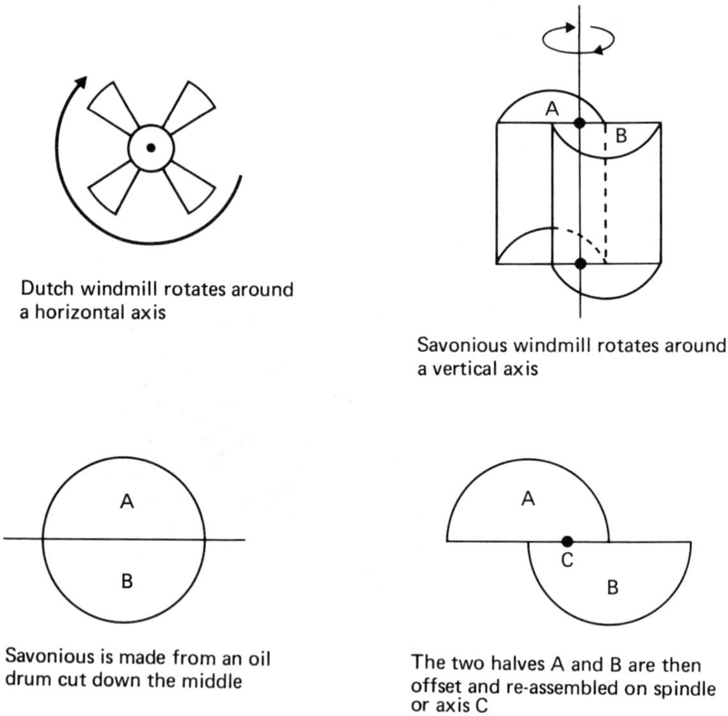

Fig. 30. The 'Savonious rotor' windmill works which ever direction the wind blows from.

water seeking the sea.

The harnessing (collection) of these great forces will be dealt with in detail in later chapters, but one delightful windmill does deserve a special mention at this point. The Savonious Rotor windmill operates in the same way as Chinese windmills of 4000 years past. It does not operate like a Dutch windmill by rotating round a horizontal axis, it rotates around a vertical axis. This has the unfortunate effect of lowering its efficiency, but it has several compensating factors, one of which is its ease of construction. By literally splitting an oil drum down the middle and offsetting the two halves onto a spindle, the 'windmill' is made. This is tremendous village technology, and Fig. 30, should make it clear.

Sunny or cloudy, day or night, Solar Energy is always arriving on earth at a rate hundreds of times faster than we are currently using our present fuels. The plethora of arrangements to collect Solar Energy has been carefully culled to give the selection you have just read, and even with the superficial first treatment they have been given, the possibilities should be clear. There are indeed many forms of Solar Energy.

Having read this far, you should be getting a good impression of the most successful methods of collecting Solar Energy. While it is natural to quickly identify your needs with a certain scheme, it would be unwise to get this preference too firmly settled in your mind. This book's objective is to help you produce a good operational system in your home. By waiting until you have seen how the other options affect your thinking, you will arrive at a sounder conclusion. The combined collection and storing schemes in a later chapter are good examples and provide some interesting alternatives.

CHAPTER 5

How to Store Solar Energy

It is an unfortunate characteristic of Solar Energy that it arrives in a quite random manner; it does not arrive at a time to suit our needs, and we cannot switch it off and on to suit our demands. In modern jargon, the input doesn't match the output; a common enough problem these days. Take, as an example, personal incomes. Here we find that batches of cash inflow are matched to random outflow by the simple expedient of adding a buffer store – the bank. By holding the inputs in store they are easily able to cope with the totally unrelated demands of the outflow. Alas, this incurs two charges.

The first is the bank charge, or, what is left behind as it passes through the store. The second is the cost of the store; equivalent to the cost of building the bank. This is not an item that normally concerns one, but an important factor in considering a Solar store. Either a store pays for its keep, or it is not worth keeping.

Consider the operation of a store. To operate correctly every store must have an input and an output; these can be two points, or one and the same point. If it is one point, then the input and output flow through the same point, rather like a store cupboard with one door for the two-way traffic. The two-way traffic is essential, but whether there are one or two doors is optional. Fig. 31 shows the two options.

Another essential feature of a store is that it should have as few losses as possible. Whatever these losses are, and wherever they occur, it is clearly desirable that the bulk of what one has put into the store one would also wish to take out. Some loss is

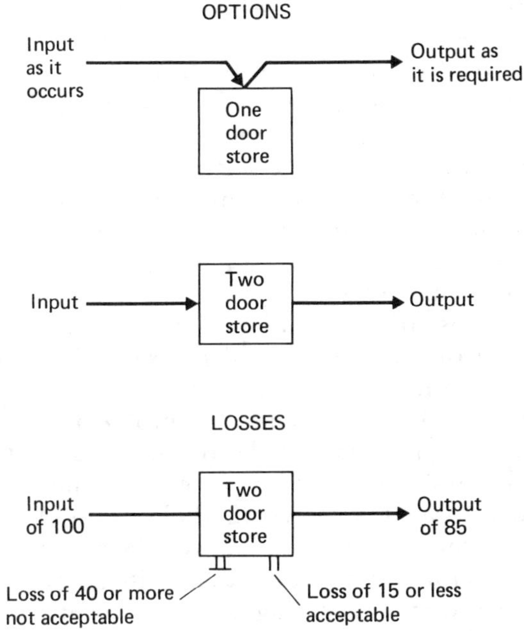

Fig. 31. The options of entry to a store. Losses from a store.

inevitable, but the objective must be to keep this as low as possible.

These two features of operation – the one- or two-door store and the minimum heat loss – describe the basic requirements of any store, whether it be to store Solar Energy or soft fruit. See Fig. 31, again.

Just as the chapter on the collection of Solar Energy took the order, heat, then light and finally movement; so we retain that order when considering the storage of Solar Energy. And again, following the collection pattern, we start with a method that stores Solar in water – a very popular idea and a two-door store.

To form the simplest store requires only that we store the collected warm water in sufficient quantity; and as long as we keep enough of it, and keep it from cooling too much, we have a perfectly good Solar heat store. (The familiar 'hot water bottle', used for bedtime comfort, is a small heat store.) As a store holds many times more water than does the collector it is connected to, there is a constant flow round and round from the collector as it gradually warms the store. Sometimes the collector receives cold water to make warm, and at other times warm water to make hot.

With the cost of the storage medium being nil (the water), the true cost of the store is hidden. Its cost is really in the container to hold the water and the heat insulating lagging around it. And, being a large container, the problem of housing it may also be considerable. As a guide to size it will be at least that of a normal hot-water tank, if one is only storing domestic water, and about 2 to 3 cubic meters (2500 liters, or 660 U.S. gallons) if one is aiming to store one or two days' home heating plus domestic water in mid-winter; that is quite a large tank.

While the word 'tank' suggests a conventional metal box there are alternatives.

Water or pebbles can be contained in a pit lined with concrete building blocks and painted inside with epoxy resin or waterproof cement. Such structures have been used as rain water stores or septic tanks for many years.

Other devices to store water or pebbles include a section of large-diameter spun concrete pipe (sewer pipe), stood on end and blocked at the bottom; or a section of 16 g. culvert pipe with one end welded up and the inside painted with epoxy resin.

'Mixing' occurs in all hot water tanks, if not guarded against, and is the undesirable cooling of the hottest water by getting mixed with the cooler water. To reduce the problem, care must be taken that the incoming cold does not disturb the hot already in the tank, and, conversely, that the incoming hot does not flow in and mix with the cold. With an indirect system as in Fig. 27, this problem is reduced, but in direct systems the points of entry of the connecting pipes are important. If you look closely at Fig. 26 you will see that the hot feed is about one-third down from the top of the tank, whereas the cold feed comes in right at the bottom. These are the best compromise positions. An additional aid that reduces the mixing problem is the inclusion of a series of flat horizontal plates inside the tank. These plates reduce any tendency towards movement of the water from top to bottom, thus helping to keep the hot where it was put and the cold where it was put.

Finally, there is the lagging to consider. This is the heat insulation around the store and associated hot pipes. The amount varies, but a thickness of perhaps 100 mm. (4 inches) gives a balance between, keeping enough heat in and not wasting money by overdoing it. If your insulation is free, you cannot overdo it, but if the lagging costs more than it saves, then you have overdone it. A good practical guide is to *feel* if the outside of the lagging is getting warm, if it is, then too much heat is getting out and more lagging is required.

We now see that our first store, the 'water store', can be described as follows. A large well-insulated container full of water, carefully piped up, and possibly fitted with internal baffle plates.

Our second practical store retains the heat not in water but in ordinary pebbles – literally a 'pebble store'. The pebbles are housed in a large bin and heat is put into this store by blowing warm air through the bin. The air does not actually penetrate the pebbles, but weaves its way through the gaps that exist between them. (It is again interesting to look back and recall that another bed warmer of years ago was a large flat stone that had been heated by the fire, and used to carry its stored heat to the bed.)

One popular version of our store consists of blocking off a half or a quarter of a cellar and filling it with pebbles. They would be small fist size for a large store, and walnut size for a smaller one. As no water is used, wooden shuttering or a brick wall is sufficient to hold the pebbles in place. Insulation is still required around the container, although it is not as import-

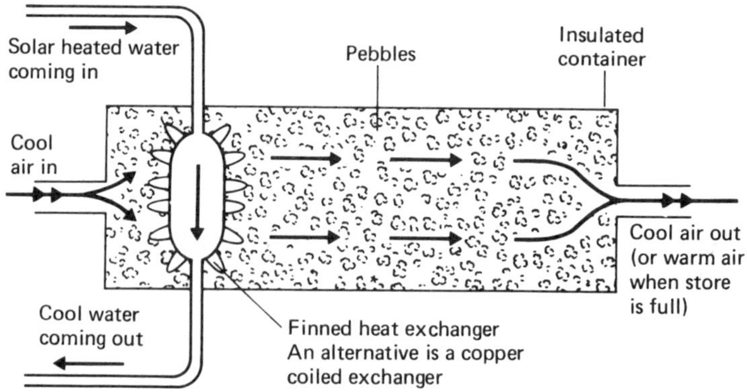

Fig. 32. How to store solar energy in a 'pebble store'. The double arrows show the filling mode; reversal of the arrows empties the store. Air-flow left to right fills the store, and air flow right to left empties again.

ant as before because 'mixing' is much less. Refer now to Fig. 32, as the detail is filled in and the operation explained.

Solar warmth is carried in at the top left by hot water in the usual manner, and, again as in the indirect system of Fig. 27, it arrives at a finned heat exchanger (other types are referred to later). But this time, instead of having the heat taken away by more water, it is removed by cool air being blown in from the left side. The air takes the heat from the fins and *carries it to the pebbles,* giving it up to them as it finds its way through the maze of passages between them. The air outlet at the right is only necessary to allow a free flow of air, and contrary to the impression given by Fig. 32, this is a one-door store. In all cases (except when the store is full) the air will come out on the right quite cool. In the case of a store half full of heat, the pebbles to the left of center are hot, and those to the right are cold. There are grades in between, but the point is that before the air gets out to the right it has given up all its heat to the pebbles (leaving those at the end cold) and emerges cool. In the case when the store is full, all the pebbles are hot and the air emerges hot and can only be put to waste.

Consider now being in the enviable position of having a full, or half-full store of Solar heat – a wholly delightful sensation as one muses on the thought of a harvest of free heat tucked away, and about as pleasant as somebody filling up your oil tank for free. How well does the heat stay where it was put? Unlike the various parts of a water store, the pebbles do not sit in intimate contact with one another. Pebbles being the rotund shape they are can only be in contact with the next one at *one small* point; thus, the heat transfer by contact along the pebbles is quite small. Consequently, 'mixing' is much reduced due to actual contact, but some does take place via little air currents from pebble to pebble and by radiation pebble to pebble. View a gradually filling pebble store as one with a steepish internal temperature gradient point that gradually moves to the right as it fills, and moves back to the left as it empties.

Getting the heat out of the pebble store, which is our final objective, means a reversal of the air flow. By blowing cool air in at the right, hot air comes out at the left to circulate round the house, underlining the fact that this really is a one-door store. If you only have a 1-day store, you will notice the reverse heat pattern coming out more than that which went in during the day, the noon heat peak coming out halfway through the evening. Short pebble stores are preferred.

If a pebble store is used with an air operated collector it should be seen as a distinct advantage that the air returned to the collector is quite cool and thus helps to achieve higher collective efficiencies. Each circuit of the air gives up nearly all its energy to the store.

Our third store is one that is only just coming into use, and should be viewed as something less proven than the previous two stores. This one is a fascinating idea that uses earth as the storage device, and you don't dig it up and put it in a bin, you use it where it is!

The 'earth store' operates by putting the heat to be stored into the ground under the house, preferably by way of a series of drain pipes buried in the earth. If the pipes are under the house, then any upward heat loss is barely a loss, but this is not an essential feature and should not limit the scheme to new houses. Fig. 33 shows the general arrangement with heat being carried into the maze of buried pipes by blowing the heat to be stored in at one end and out at the other. As the hot air goes through the pipes, it gives up its warmth to the walls of the pipes which in turn gives it up to heat the surrounding earth. (Porous, land drainage pipes give up the heat better but can

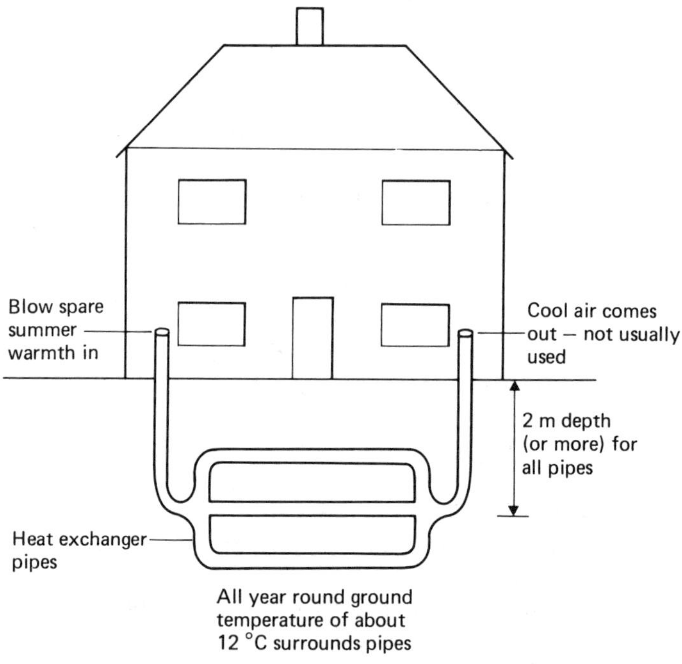

Fig. 33. How to store spare summer heat for use in the winter – and you may get back more than you put in! Reversal of the summer input flow brings the warmth out.

[Note. These stores cannot be considered if the pipes are on or below the water table. If they were the cross flow would carry all the stored heat away.]

only be used on dry land; otherwise they fill with water.) The earth next to the pipes and the pipes themselves, surrounded as they are by more earth, hold most of the heat until it is required. Recovery of the heat is achieved by reversing the air flow through the pipes.

The earth store has three unique features that make it such an attractive and tempting proposition. One, it takes up no room in the house. Two, it can be used as an inter-season store – put heat in in the summer and take it out in the winter. Three: the store can pay you interest by giving out more than you put in! The legendary something for nothing. Look at it this way; the ground beneath our feet always has some warmth in it, and at a depth of two meters (6½ ft.) is typically 12°C (54°F) – a constant temperature. No doubt you can recall visiting some cave or other and being told by the guide, "it's 15°C (59°F) down here winter and summer"; it's the same reason. Thus, we can see that we are putting our heat into a store surrounded on all sides by an immense blanket of warm insulation that starts off at 12°C (54°F) *or more.* (As in the ideal case, the house effectively buries the pipes further in the ground, it is a reasonable assumption that 2 meters under the house would be hotter than 2 meters down in an open field.)

But, to temper the euphoria just created, there may be a disadvantage. While there is plenty of warm air coming out of the store, it may not be quite warm enough. An 18°C (64°F) air blown about the house has a decided chilling effect (recall, how, if you fan yourself with the air of a hot room it still feels cool). Consequently, at least 25°C (77°F) air has to be used. This means that if the heat from the earth store is not hot enough it has one of three uses. One, it can be used as background warmth for the house. Two, it can be put through a heat pump to boost it up before being used in the house as primary heat. Three, it can be used as low-grade heat for a swimming pool or greenhouse.

As a further encouragement to would-be earth store users we know of at least two successful installations in the northern U.S. State, Connecticut, where the houses are heated using only the warmth from the ground! A single pipe 600 mm (2 ft.) in diameter forms a large loop round the grounds at 2 meters (6½ ft.) depth, and sufficient warmth is extracted from it to heat the whole house. To get the heat up to a useable temperature they do have to use a heat pump, but remember they didn't put *any* heat into the ground.

Another new and intriguing idea coming into favor is to use a pebble store in the ground. By doing this, not only is the container free but also the insulation! Two versions are shown in Fig. 34. In version A, the air flow is the same as the previous pebble store, whereas in version B, the flow is, down from the top to fill the store, and back up to the top to empty it.

A fifth store is a stimulated pebble store cleverly formed by stacking 5-liter (1 gal.) plastic bottles filled with water. Spaces are left between the bottles to allow air to weave its way between them as if they were large pebbles. Note that many of the advantages of the pebble store have been retained, plus getting the much greater heat storage capacity of the water without its 'mixing' problem.

After going through these five storage schemes it is apparent that the Solar fraternity have a sheer genius for coming up with 'no cost' materials. The frugality of water, pebbles and earth is not easily matched, in sharp contrast to the cost of uranium, plutonium and godknowswhatium.

Our sixth store is quite different from the previous five; it still takes in and gives back heat, but its operation depends on the phenomenon which physicists call 'a change of state'. Our previous five examples showed no change of state as we put more heat in, they just became hotter, but at no time did they become anything but water, pebbles or earth. A common example of a change of state heat store is water changing to steam, and while we are not using water in our sixth store, it serves well to explain the phenomenon.

The advantage of using a change of state method is that at the actual change point, the material in changing absorbs very large amounts of heat without raising its temperature. This energy is actually taken into store in the material, not to make

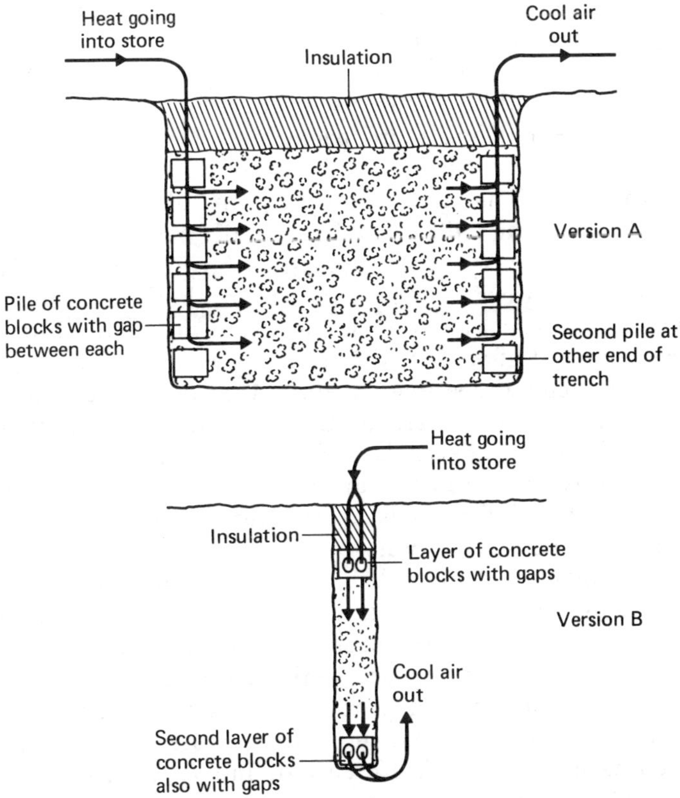

Fig. 34. An 'earth pebble' store. In both versions a 300 mm. (12 in.) slot is cut in the ground and filled with pebbles. Heat passes in to the pebbles via gaps between hollow concrete blocks. Reverse flow recovers the heat from the *same point.*

it hotter, but to change its physical state from water to steam. That it does not rise above its boiling point of 100°C is not a disadvantage, quite the reverse; it takes in and stores more heat without getting hotter. Once it has taken in all it can during the change, the water (now steam) can take in further heat and rise above its boiling point of 100°C. This store takes in some warmth to rise to 100°C; more warmth to change to steam; and even more warmth to go above 100°C.

In a practical change of state store, crystals such as Glauber Salts (with some additives) are used. Thus the name, the 'chemical store'. The chemicals are packed into long plastic tubes, rather like fluorescent lamps, and stacked so that warm air can be blown right along the outside of each tube. The warm air eventually heats up the chemicals until they melt at their change point. This type of store is the most compact there is, and it is due entirely to the phenomenon; they can absorb ten times as much heat as a comparable sized water store but they do have a limited life before the chemicals have to be changed.

This completes the storage of heat. We now proceed to consider storing the light part of the Solar rays. Our first light collector, the wizard-like Solar cells produce electricity that can be fed directly into ordinary automobile batteries. By the mere connection of sufficient cells to make up 12 volts (each cell is about ½ volt) the work is done.

Our second use of Solar light was the photobiological process, but, as yet, nobody has a practical scheme using these principles. One of the most promising of several being worked on is a cell that when exposed to Solar produces hydrogen. If that cell gets efficient enough to come out of the lab, then we merely collect and compress the hydrogen into small cylinders. We then have Solar that is stored indefinitely, and, stored in a transportable form. Without departing too far from our brief, it is a lovely thought to recall that a standard car engine is readily converted to run on hydrogen; and as a bonus the waste products from the engine are not poisonous fumes but water! The students at Loughborough College, England, who designed and produced such a car proved the point by drinking the exhaust outlet.

And the third light user was the photobiological process, our ever-active friends the plants. Their collection and store of Solar Energy goes on all the time, with scientists working to reduce the time taken before the plants produce a useable product. But storage, in the sense we are considering it in this chapter, is only encompassed by the storage of methane gas from decomposed vegetation. This we discuss in a later chapter, when the complete process of production and store, prior to use, is covered.

Returning briefly to the way that the plants do store up the energy, it is worth noting that the period of their store is inter-seasonal. When we eat fruit or vegetables, we are recovering energy in the summer/winter that they took into store in the spring/summer. And, in the case of burning wood or methane gas, we are taking Solar Energy out of nature's store anywhere from one to several years after it was put in.

Observe please: in the heat stores we are always trying to *extend* the storage time before we recover the energy, yet with the photobiological stores we are looking for techniques that speed up the process and thus *reduce* the storage time.

Finally, we look at the storage of Solar when it comes to us as movement of air and water. As all our examples produce rotary motion, we will consider wind and water as one, in the four practical methods discussed.

The first, and by far the most popular method, is just to drive an electric generator and again store the electricity in batteries. There is no need for these to be separate batteries if you are already filling some from Solar cells, but be careful when connecting them up that the generator does not damage the Solar cells. (An isolating diode in each feed to the battery appears to be called for here.)

Second, consider using the output of the generator (often 12 volts) to boost your hot water, 12-volt immersion heaters are available from car or camp accessory shops. By effectively storing your own electricity *to offset electricity* from the utility company, you make the best possible use of the generator. That is first based on the near certainty that electricity will be the most expensive form of energy in your area. And second, that you have not put energy into a store, then taken it out, and then used it; you use the energy *from the store* you put it in. That way you avoid one set of losses. Remember also, that the output from an alternator (which supplies A.C. instead of the usual D.C. from a generator) can be fed straight to an immersion heater. It is not necessary to change it to D.C. as one would if storing it in a battery. Thus the losses of that change are also avoided.

The third-store method merely rotates a water pump, the idea being to lift water up and store it at some high point. This is either a large tank, or more likely, a pond on high ground. Upon release, the downward flow of the water either does useful work, or produces electricity to recover the stored energy. To derive much benefit from the scheme requires considering it on a fairly large scale, plus accepting losses of about 30%. For example, to store our usual unit of one kilowatt hour (kw/hr), requires that 47 cubic meters (47,000 liters or 12,400 U.S. gals.) of water is lifted up a height of 10 meters (33 ft.).

Our fourth store uses air as its storage medium, the idea being to pump or compress air into a large container and release it to recover the energy stored. The Electrical Institute of Research in Quebec is presently carrying out a study to use a cave in the Gulf of St. Lawrence area as its container, and a second study is under way in Germany. On a more practical level, 'container' refers to a tubular tank specially prepared and strengthened to withstand the intended pressure.

These last two schemes will not appeal to many readers, but they do work. So, if you have a windmill still gaily galloping round after having filled up all your batteries, you may consider using your extra surplus to operate scheme 3 or 4.

Thus, we have at least five methods of storing Solar heat, at least one of storing Solar light and four for storing wind/water power. There are several other possibilities, but few if any are more than enthusiastic gleams in inventors' eyes, or at best, carefully nurtured lab models. Some of them will win through and provide alternatives in the future, but at the moment, the ten schemes we have gone through all work and have been proven in practical schemes around the world.

With a dozen or so possible collectors and now ten stores to choose from the number of combinations abounds. Think them through to see how they can be adapted to your needs. Think about using two different types of collectors – perhaps, a simple one at the start feeding into a better one at the finish – or consider two different stores to feed into. You have some very good components to work with; they should make up into fine schemes to suit your particular needs.

Our *specific* schemes follow in later chapters.

CHAPTER 6

Combined Collector-Stores — Passive Systems

The bringing together of the collector and the store to form a single component thrusts upon us a whole new concept of Solar Energy. That sounds like an extraordinary statement, but these systems are so different from what we have seen before that they could eventually dominate the Solar world. To give them the identity they deserve, they are separately classified as 'passive systems'. Thus far we have only discussed 'active systems', systems where there is near constant activity and interchange of heat between the main parts of the system, and that is generally due to the action of a motor. Passive systems are characterized by minimal motor activity, and what little there is is normally limited to the changeover period at the start and finish of the Solar day.

With little fear of contradiction, certainly in the long term, it can be stated that passive systems offer the best possibilities for Solar utilization. The reason for the qualification 'in the long term' is that passive systems are more suited to new buildings. Fortunately, this does not exclude them from being considered, in part, for existing houses. This you will see as we proceed.

Three practical examples that we now look at epitomize these passive systems, and lead us on to the more general possibilities. First, is the scheme invented by a far-sighted English architect, one A. E. Morgan, and the school extension he designed in mid-England using a giant glass wall and thick concrete floors. Second, we look at a scheme invented by Harold Hay that he tried out in Arizona and California, using as the basis, shallow roof ponds full of water. Third, we look at the scheme invented by Thrombé and Michel and built many times in various parts of France. They opt for an arrangement based on a heavy concrete or brick wall with a circulating air current. After the description of the three schemes we look at several other passive schemes which are on a less and less grand scale. Eventually it will be seen what is dawning on the Solar fraternity, and indeed, the world at large.

Passive schemes offer an extraordinary breadth of Solar possibilities.

We make no excuse for this chapter traversing the apparent reverse route from specific schemes to general principles. It is used to deliberately reflect the actual course of events.

In 1959, architect Morgan trod a lonely path as he pioneered what was to become Europe's most noted Solar building, the addition to St. George's School at Wallasey, Cheshire, U.K.

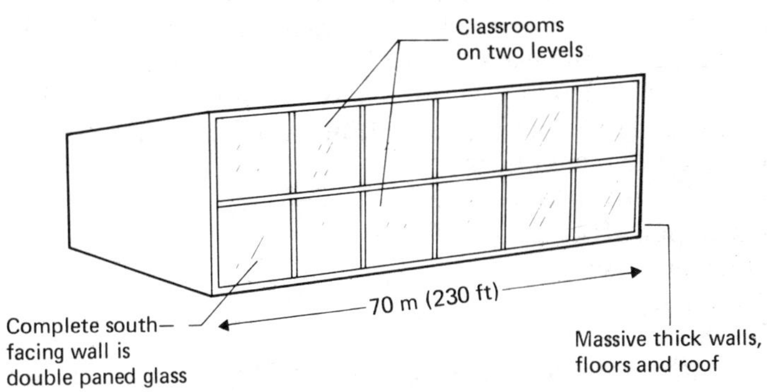

Fig. 35. Europe's most famous Solar building, the extension to St. George's School, Wallasey, Cheshire, U.K.

Fig. 35 shows this 70-meter (230-ft.) long building with its double-glass south facing wall. The glass wall is one giant collector or input for the store, and the store is the thick dense construction of the walls and floor. That is basically it! Great attention has been paid to the heat insulation of the building, and to the exclusion of drafts, but with the aid of these not unusual precautions the school is kept to a very comfortable temperature with no apparent heating arrangements. Actually, the warmth from the lights and the energetic little bodies of its occupants provide some heat, but the remainder is Solar heat. This is how it works. Solar radiation comes in through the glass wall and either strikes the thick walls or the thick floors, gradually heating them up as the heat is taken into store. On sunless days the heat comes out from the floors and walls to heat the occupants. The feature of the building climate is its steadiness. By absorbing much of the Sun when it appears, overheated classrooms are avoided, and by returning the heat when it is not out, the reverse also holds true. The general massiveness of the construction is itself responsible for damping down the temperature fluctuations and the glass wall for the actual collecting of the Solar in the first place. Clearly it is the structure itself that is the collector and the store, and it stores enough to give a one-week carry-through of heat. The school has been in routine use since 1961, giving a close approximation to 100% Solar heating without any apparent Solar equipment at a site as far North as Edmonton, Alberta! Further information can be obtained from;

The Director of Development
Municipal Offices,
Brighton Street,
Wallasey,
Cheshire, Great Britain.

So to the second passive scheme. In 1967 two other pioneer inventors, one in the old world, one in the new, started to design radically different Solar houses.

In the latter case, Harold Hay set to work in Arizona on a system that uses water to collect and store the heat. This is the same water for both functions and it is static water; that is, it is not water that collects heat at one point and then moves to store it at another. It all happens right there where the water lies. Fig. 36 shows the beautiful simplicity of Hay's house, and this is how it works.

The Figure shows the house has a roof pond with moveable covers. These covers are shown stored in their winter daytime position over the car port, thus enabling the roof pond to collect Solar heat. As each succeeding day's heat is taken in, the

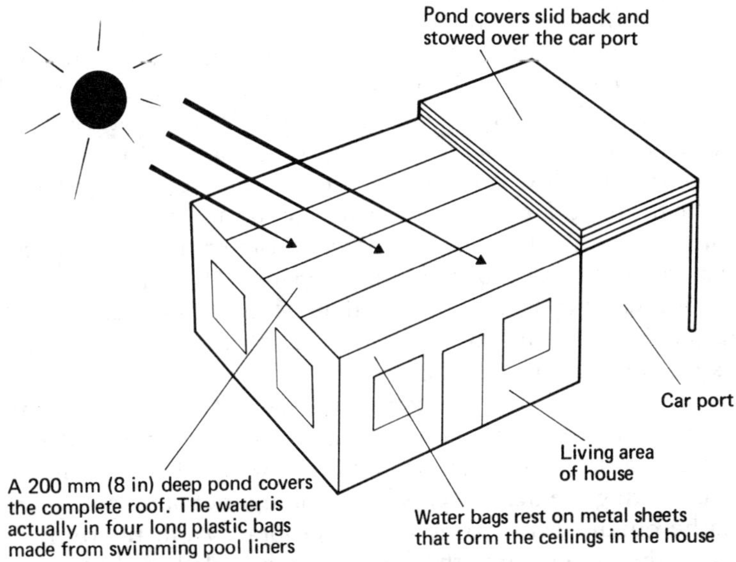

Fig. 36. Harold Hay's house has a shallow roof pond. Movable shutters let the Sun warm the pond by day, and the covers slid on at night keep the warmth in to radiate down into the house.

pond gradually warms up and at the end of each day the heat insulating covers are slid over the pond thus both saving upward waste, and enhancing the flow of downward heat. The downward flow into the house takes place simply because the pond is resting on a metal plate that forms the ceiling of each and every room in the house. This 100% coverage of the house provides a gentle overall warmth that eliminates hot or cold spots and violent under or overshoots of room temperature. If the outside temperature did shoot up, the pond and its metal ceiling sheet would only change a few degrees in a day: similarly if the outside temperature dropped down dramatically. The massive control by the structure (and this time it is an engineered massiveness in water) irons out the variations. Careful independent tests carried out over several years speak in glowing terms of the adequacy and the quality of warmth. This is always a feature of passive systems as the source of heat is at a *low* temperature. It is not derived from some very hot heat source and injected into the house to spread out and mix with the other air. It is an all-pervading warmth from a source a mere few degrees hotter than the desired room temperature.

As Harold Hay lives in the hot Southwest, he also needs cooling. (It still comes as a surprise to many people in the world to hear the U.S. electric power crises come in the summer, because of the air, conditioning demand. Also, that the utility bill for cooling can be more than is spent on heating.) But Harold Hay defeats this cost also; all he does is to reverse his system for the summer months.

On summer nights the covers are slid off of the roof so that the night air can cool the pond and make it ready to radiate its 'coolth' (a delightful new self-description word) down into the house during the next day. Additional night cooling of the pond can be obtained by letting water evaporate from the roof. This is not the water from the bags (that is sealed in) but is extra water over the bags and allowed to evaporate. (Cooling by evaporation is how our bodies keep cool as they evaporate sweat, and why when we emerge from even a hot shower we feel cool until all the surface water has been removed.) In the day the covers are slid back over the pond to keep the Sun off of the roof, so letting the night-generated coolness radiate down into the house. Hay achieves 100% Solar heating and 100% Solar cooling. Marvellous.

The above method of cooling will not be so effective, or for that matter so necessary in temperate climates, but the heat-collecting/storing part is well worth thinking about to provide some, if not all, of the house heat.

One improvement added to the basic bag of water forming

the pond is to stimulate a double 'glass' layer effect by fitting a second sheet of plastic on the Sun side of the bag (the top). The two layers are then inflated and kept apart with light air pressure, so forming a thick protective air layer over the complete pond. The Sun can still get through to the water but the heat of the water cannot escape as well from the topside of the pond. For night cooling, the air layer is generally let out. For more information write:

Skytherm Processes & Engineering
2424 Wilshire Boulevard
Los Angeles, California, 90067, U.S.A.

The second house is shown in Fig. 37, and is the basic arrangement worked out by Thrombé and Michel. Again it is a very simple system: virtually nothing moves except the warm air going round the house. There is neither water, nor pebbles, nor earth in this system; it is based on air slowly revolving round a heavy concrete or brick wall. The wall serves as both collector and heat store. Let us now see how this third intriguing scheme heats a house. Fig. 37 shows the detail.

The incoming Sun rays go through the double-glass cover to heat up the blackened concrete wall; our easily recognizable 'greenhouse effect' and 'blackened surface' are well in evidence here. But in this installation, the layer of warm air created between the inner glass and the wall is allowed to flow up and out; just like hot air in a chimney flows up to get out. Note that the so-called 'chimney' is only 100 mm. (4 in.) across but stretches the full width of the house. Cooler air is then sucked into the chimney from the room, at near-floor level, completing the air flow circuit round the concrete wall. To recap: the air circuit is – cool air in at the bottom, up the thin wide chimney picking up warmth from the wall, and round at the top and back into the room as warm air.

With a small alteration to the system a simplistic cooling arrangement can also be put into effect for the warmer weather. By letting the warm air escape out of the top of the chimney at A in Fig. 37, instead of entering the room, not only is the hot air *not* warming up the room in the summer, but it also draws

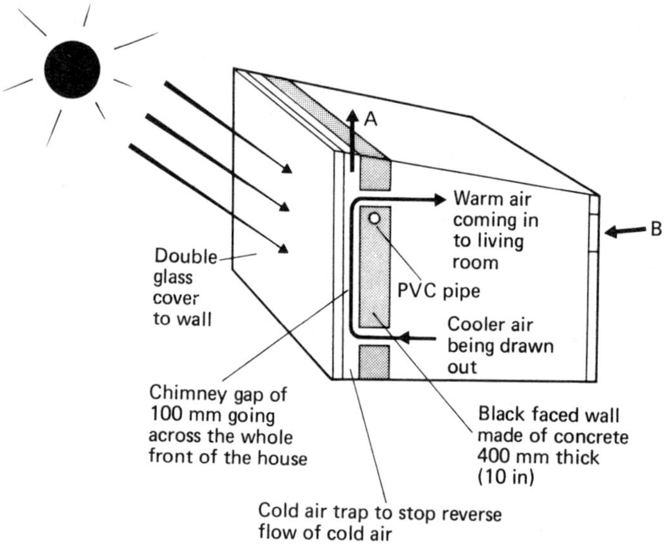

Fig. 37. Prof. Thrombé Michel's houses have a thick south-facing wall behind a glass heat trap. The heat is stored in the wall, and this creates a flow of warm air that circulates round the house.

warm air from the house. By then allowing the cooler air to enter the house at B in Fig. 37, a cross flow of air is created to cool, or at least ventilate, the house. As the air inlet vent B is at the rear of the house, and out of the Sun, it is the coolest air one can find. Again, marvellous.

A feature that another innovator has added to the 'wall' provides Solar pre-heat for domestic hot water. The incoming cold water is simply fed via a PVC pipe that has been embedded across the front of the wall, the hottest point.

For a 'wall' house in the more northerly U.S. states (lat. 42°) a relationship has been established to help decide the area of wall required. For a poorly insulated house the area should be one-sixth of the volume being heated, and for an average insulated house one-tenth. That is, one square meter heats ten cubic meters.

As a very general categorization of the last two houses; note that as the Hay scheme is on a roof, this will operate best in areas nearer the equator due to the higher angle that the Sun rises (the two houses were built at about lat. 34°, i.e. at the top end of band 2). Whereas, the Thrombé/Michel scheme, collecting from a wall, works better in the higher latitudes, i.e. band 3).

So we have two very effective house schemes, one on a roof and one on a wall, one with water and the other with concrete. We will now see how a whole range of passive schemes develop from these two, without forgetting the ideas used in the school.

A third equally well-known house scheme is that designed by young Steve Baer – entrepreneur extraodinary – and his oil drum home in Albuquerque. (Steve is the sort of character who says, 'Engineers are about the last people to let loose on this Solar stuff as they have too much to answer for already considering the other technologies they have gotten us tied up with. Solar needs 'intuitive engineers', if you can't feel its simplicity and affinity to nature it's not likely you're going to

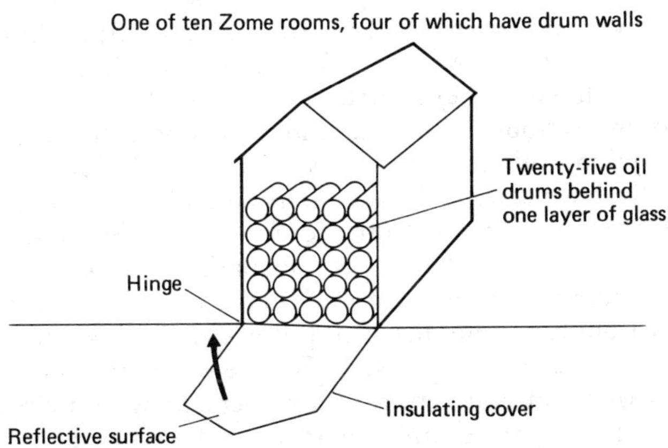

Fig. 38. Part of Steve Baer's house complex. A 'Zome' is the name given to these many-sided rooms.

make a good job of Solar.' His home reflects Steve, see Fig. 38.

This is how it works. The wall of water-filled 55 gallon oil drums are his collector and his store; by painting them black and covering the outside with glass, they absorb much of the day's heat. This heat comes partially from direct Sunlight and partially from the ground reflector. At the end of the Solar day, the reflector, which is also an insulator, is swung up to cover the ends of the drums. The heat loss to the outside is stopped (and to some extent reflected inwards) and the flow of warmth to the house increased. If not exactly an elegant scheme, it certainly is simple (which is a form of elegance) and it is clearly a passive scheme. Further information on the house and other Steve Baer Solar devices can be obtained by send $*1.00 to:*

Zomeworks, Inc.,
Albuquerque, New Mexico, U.S.A.

By comparing the three houses of Hay, Thrombé/Michel and Baer, we arrive at an interesting conclusion.

As we have seen, Hay's house with its roof collector is more suited to bands 1 and 2 operation, whereas Trombé/Michel with its wall collector is more suited to band 3; it thus seems logical that Baer's house with its water wall should really be grouped with Thrombé/Michel. (The fact that he used a ground reflector at his home in band 2 tends to endorse the suggestion.) And by taking this through to its logical conclusion we deduce that as we have a wall of water scheme and a wall of concrete scheme, the complement to the roof of water scheme should be a roof of concrete scheme, as the basis of a fourth passive system. While we are not aware of such a scheme it is a 'natural' that must appear sometime if not already. The Wallasey school has flat floor surfaces as the store but we were thinking more on the lines of a flat concrete roof arrangement with a circulating air flow. In comparing concrete or water these figures should be useful.

For the same *volume* (assuming no gaps):

Water stores 100 heat units – to concrete 61, to bricks 45.

For the same *weight:*

Water stores 100 heat units – to concretes 27, to bricks 20.

By arranging the three previous schemes into a pattern we have evolved a fourth possible scheme, if we now take the Steve Baer reflector and add it to the others, we see that the four now becomes eight. Using ground reflectors to improve the collecting power of concrete or water walls, and then doubling as night insulators provide few problems, but the reflector/insulator sticking up vertically from a roof is somewhat more tricky. Damage from wind is one problem and architectural unacceptability is another. Both are minimized by making the collector area long and narrow on an East–West line. That way the reflector is only as high as the narrow dimension. If reflectors are fitted, they improve the apparent efficiency of the collector, and also make the four additional schemes suitable for use in any Band. If doubling as insulators, they also cut down the night heat loss in the winter and the unwanted day heat gain in the Summer.

Weighing in with his own passive system is the redoubtable Harry Thomason and a version of a roof pond . . . a complete break from his usual thoughts. He calls it his 'Sunny South' model and it has one particular feature to interest us. See Fig. 39.

While the roof pond, plus air bubble, plus reflector have been discussed, the following idea is new. Water that has been lying on the roof all day collecting heat is, at the end of the Solar day, allowed to drain from the roof and fill up a similar space under the house, this way the heat can rise up to warm the occupants. That in itself is clever enough – but by filling the under-floor space with pebbles (which the water permeates), the heat from the water is transferred to them. In the morning the water is pumped back to refill the roof pond, there to spend its day dutifully collecting heat to take back to the store in the evening.

There is no major objection to applying the same principle, of moving water night and morning, to a scheme where the water is contained in a wall and not on a roof.

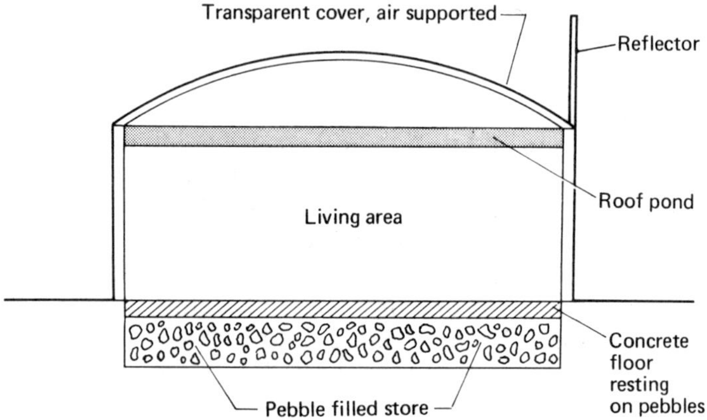

Fig. 39. The Thomason 'Sunny South' house. The water on the roof, after collecting heat all day, is transferred to the pebble store at night where it gives up its days harvest before being returned to the roof in the morning to resume collecting.

With ten possible passive schemes we now turn our attention to the *cover* for the collector/store area. Thus far we have only considered the attitude of the collector (horizontal or vertical), the material it is made of (solid or liquid), and the possible addition of a reflector to generate greater input plus freedom of geographical location.

The cover for the collector/store has many alternatives; it can either be *trapping* Sunlight behind a transparent screen (glass or plastic, *reflecting* more Sun in or reflecting collected heat back into the store; or acting as a heat *insulator* to again keep the heat in the store.

As with the prior examination of the three houses, we will first look at some specific examples and then see if we can form them into a logical pattern.

Fig. 40 shows additions to the Thrombé/Michel wall. As an alternative to the cooling effect described earlier, a shade can be inserted between the inner glass pane and the wall to achieve a similar objective. The shade is reflective on the Sun side to

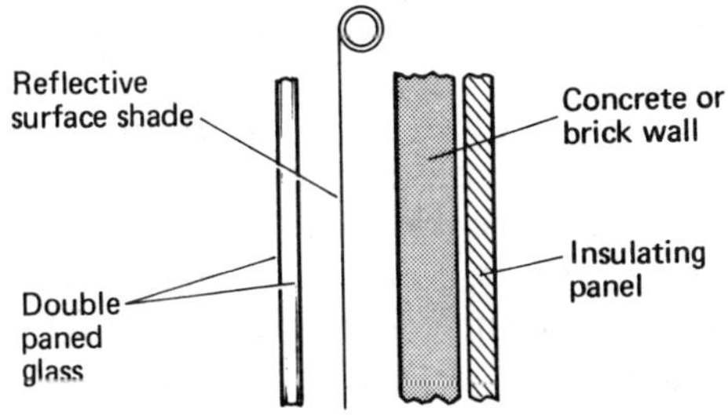

Fig. 40. A shade can be added to the front of the Thrombé/Michel wall, and an insulating panel at the rear.

reject the incoming Solar. A further addition could be the insulating panel also shown in the Figure. This panel is added as required, to accept or reject the heat that soaks through the wall into the house. A typical soak-through period would be 8 hours; that is the inside of the wall gets to its maximum temperature at 8 p.m.

The right side of Fig. 41 shows the use of louvres as a shade. In the arrangement shown, the louvres are made reflective so that when closed they form a complete mirror as on the left of the diagram. The various modes are: A. Summer day-Solar input is rejected; B. Winter night-stored Solar is conserved; C.

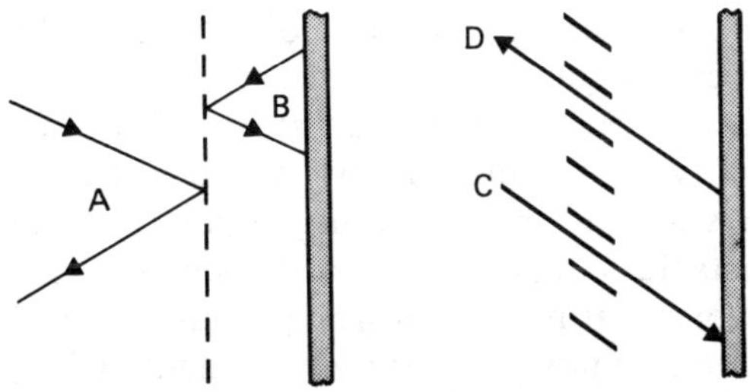

Fig. 41. Louvres added to a 'wall' but they could equally as well be added to a roof. A is summer day and D is summer night, C is winter day and B is winter night.

Winter day-Solar is allowed to enter; D. Summer night-heat is allowed out.

Fig. 42, although double-paned glass does reduce heat loss there is still a considerable loss due to circulating air currents *inside* the two panes. Steve Baer's idea is to fill the air space with polystyrene pellets (which are mainly air) so completely stopping the air currents. The pellets are blown in at, say, night-time and sucked out in the morning. It's great fun to watch, and it really works well.

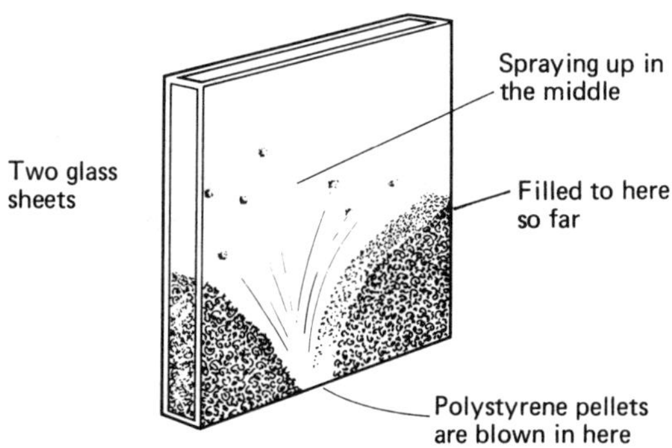

Fig. 42. The 'Beadwall' idea. By filling the space between the glass sheets, heat losses are further reduced. A very neat form of 'moveable' insulation.

Fig. 43, by the simple addition of a plastic sheet, an insulating air layer can be added to a heat store area. The layer is most effective when gathering and storing heat, and can easily be removed (by extracting the air) when not required – such as when one is summer night cooling. Again, there appears to be no major obstacle to adapting the arrangement for use on a wall scheme. As a general note of warning; plastic film hardens and cracks after long exposure to direct Sunlight. Ultra-violet resistant grades are now available and lengthen the expected life from 1 to 5 or 10 years. There is little problem with the normal grade if used behind glass.

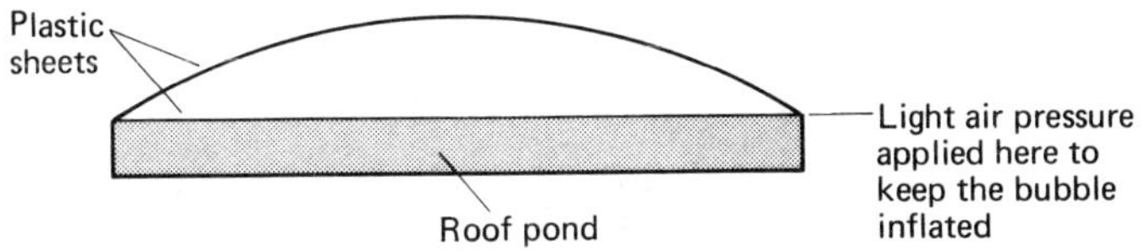

43. The inflatable air layer over the Hay roof pond. Another neat form of moveable insulation.

In Fig. 44 the absorber plate is merely a thin metal plate painted black; it converts the Solar rays to warm air and so creates an upward air flow between itself and the glass. If the warm air is then directed into a false ceiling, or duct, the air can be taken to other parts of the house. The most useful version of the absorber plate is made by sandwiching a rigid foam insulating panel with reflective foil on one side and a black sheet on the other. This sandwich will then serve winter and summer (see Figs. 44 and 45) plus achieving a reduction of the heat losses from the room at night.

A small U.S. Corporation, Kalwall Corp., is doing much

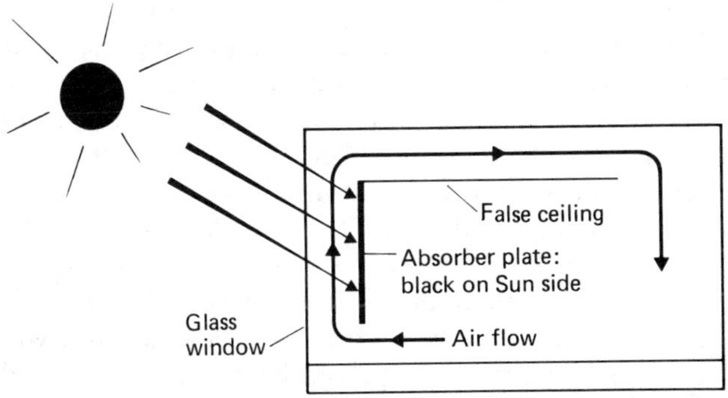

Fig. 44. Winter. By hanging an absorber plate over a window, A Thrombé Michel type chimney is formed.

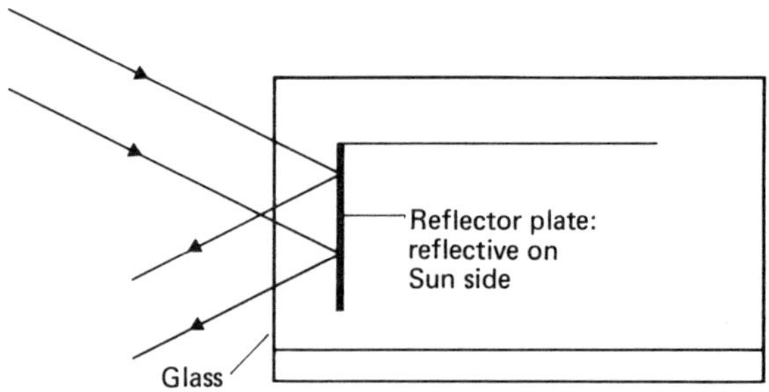

Fig. 45. Summer. The reflective side now rejects the solar.

research into materials for passive systems, and one of their areas of activity has been to develop Steve Baer's drum wall. They now offer cylindrical tubes in two diameters, to replace the drums, made from reinforced fiberglass. They can be obtained with open or closed end in 300 mm. (12-in.), or 450 mm. (18-in.) diameter up to 3 meters (10 ft.) long in various gay colors.

The idea, as shown in Fig. 46, is to make walls of water-filled tubes by standing the tubes on end. (The tubes are also quite suited for filling with pebbles or chemicals.) The tube wall is normally sited behind a transparent sheet and boxed in at the back. The rear half of the tubes being painted black to intercept the Solar rays. Walls such as this have been built into South facing walls of houses, and in one instance the "Beadwall" idea was added to it to fill up the space between the tubes and the transparent sheet to hold the heat in at night.

A more composite version of the tube wall idea is a box-like device called a Solar battery. The battery comprises a transparent plastic box with an array of water-filled plastic tubes and *moveable* insulation panels that go between the box walls and the tubes. The battery is completed by having its own air inlet and outlet for easy connection to the house system. Kalwall Corp. are located at 111 Candia Road, Manchester, New Hampshire, 03103, U.S.A.

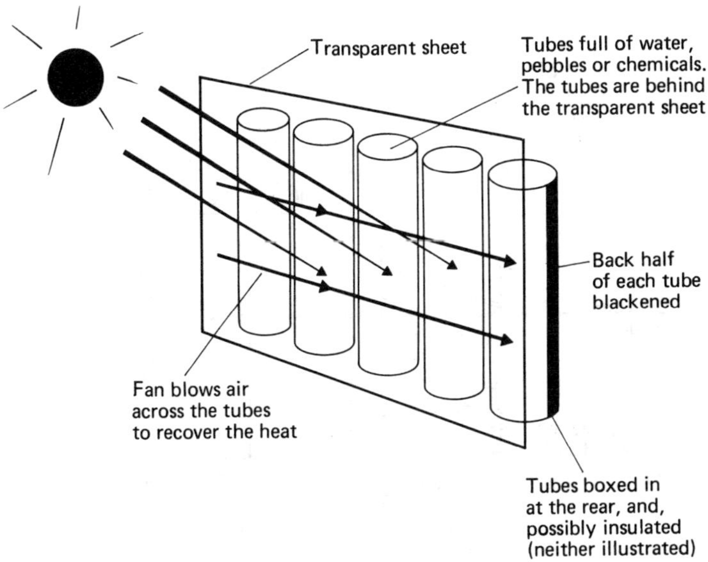

Fig. 46. The tube wall of Kalwall. The 'transparent' front sheet is often translucent plastic thus giving a quite pleasant appearance.

Fig. 47 shows a ready-made easily fitted device that is intended to improve the Solar performance of an ordinary window. Some of the techniques used are already familiar to us, but it is a neat piece of engineering and worthy of discussion. The unit is basically three rolled shades that can be individually pulled down over a window. The shades slide in relatively airtight channels which are an integral part of the fitment. As drawn, the unit is in its 'heating and insulating mode' with the middle and rear (furthest from the glass) shades pulled down. Consider first the heating action. Two separate air layers have now been formed behind the window; the first is between the middle shade and the glass and is a layer that will suffer some leakage due to the normal cracks in the window fit; the second is a fully controlled heat chamber *between* the two shades. This chamber is encompassed by the middle shade which is clear, and the rear

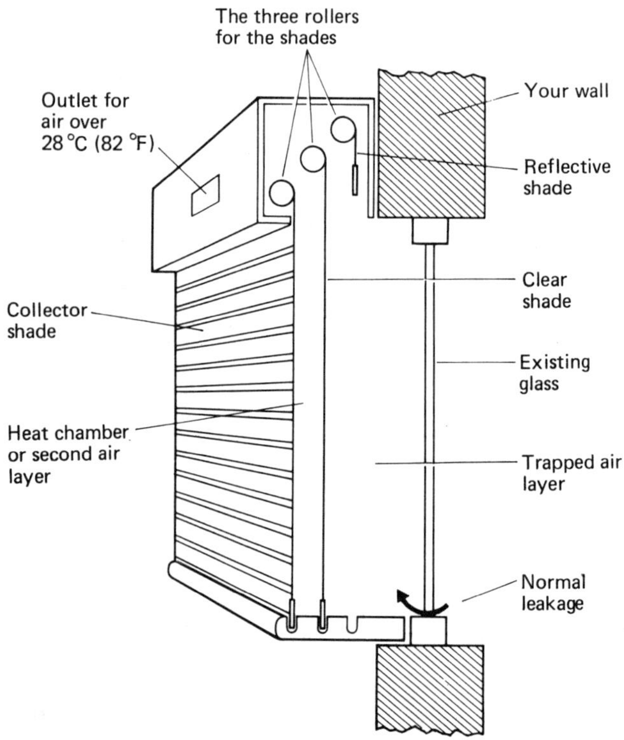

Fig. 47. The 'Insealshaid' device in its 'Heating and Insulating Mode'. Amongst other advantages, the room now has presented to it a warm area (the back of the heat chamber) instead of a cold sheet of glass.

shade which is one-third clear and two-thirds collective strips. In-coming Solar is (mainly) intercepted by the strips so warming up the heat chamber. The trapped hot air in the chamber quickly builds up temperature until it exceeds 28°C (82°F), when it is released into the room. The idea is to present to the room a warm area, the back of the chamber, instead of the cold window. The unit cannot have any magical way of collecting more Solar Energy than comes in through the window. It can merely let in an amount close to that which is arriving. What it can, and does do, is to greatly reduce the heat loss going *out*. This is achieved by:

(a) The heat chamber creating a warm surface (the room side of the collector shade), thereby cutting down the conduction losses relative to the previous loss through a cold windowpane.

(b) Eliminating all incoming cold air and all out-going warm air, by 100% sealing of the window and its frame.

(c) Presenting a reflective surface on the room side of the collector shade, thus reducing radiant heat losses by reflecting them back into the room.

(d) Greatly reducing the thermal loss, or U factor, through the window. If a normal single pane window loses 100 heat units, a double-pane unit loses 53 and with the two shades down this unit loses 27 units. At night the third shade can be pulled and then the loss is 16 units!

When in the 'cooling mode', the middle and front (reflective) shades only are used giving this time:

(a) 80% of the Solar heat reflected away.

(b) Two protective air layers to keep out the heat.

(c) Greatly reduced infiltration of hot air via crackage.

One great thing about this Solar device is that it keeps on working and earning its keep 24 hours a day all year round. Further information from:

Arkticseal Systems, Inc.
P.O. Box 428,
Butler, Wisconsin, 53007, U.S.A.

Having now been through about ten 'oddment' passive schemes, it is possible to arrive at some generalizations, if not an exact pattern.

Three objectives of the covers have been clear from the outset – insulate, reflect and trap (sunlight). And the methods we have seen used to achieve them include:

Insulate	Reflect	Trap
Plastic air bubble	Plastic, mirrored	Air layer
Trapped air layer	Glass, mirrored	Water layer
Beadwell		Metal Plate
Flexible sheet		
Rigid sheet		

The permutations and possibilities with the above abound. Further points that can now be appreciated are:

(1) Moveable barriers (insulators, reflectors or traps) ensure that the best of the winter *and* the summer conditions are obtained.

(2) Dual use of barriers (black one side, reflective the other) ensures better utilization of investment.

(3) Dual use of equipment (winter and summer) ensures better utilization of investment.

(4) A cocoon of fixed insulation, while keeping heat in, also keeps Solar heat out.

(5) Techniques designed for roof schemes can often be transplanted to wall schemes and in reverse.

(6) Bulky structures give better temperature stability, whereas windows (without bulk in the building) give the opposite.

(7) Insulation placed on the outside of the walls keeps the temperature stabilizing effect of their mass *inside* the living area.

Passive schemes offer to those contemplating house construction many complete schemes and possibilities that are difficult to ignore. Passive schemes also offer to those with homes many possibilities, although they cannot be applied on such a large scale. Look carefully at your situation and you will see some way that it can be improved. Look at your South wall and say, 'there is the best part of one kilowatt (kw.) arriving on every square meter (yard) for maybe 2000 hours a year. There must be some alteration I can make to get some of that working for me! Simple alteration, using the passive systems we have just been through, provide benefits that will pay off for evermore.

CHAPTER 7

How to Use Solar Energy

By implementing the previous chapters we could now have a supply of Solar heat coming from our collector, and a possible further supply ready to draw from our store. In this chapter we first examine all the energy forms discussed and arrange them in logical order; this way it is easier to see if the most logical sequence of operation exists. But the second (and by far the larger) part of the chapter examines the structure of the abode in which the energy is used. By understanding the gains and losses of a building, maximum benefit is obtained from the energy used in that building – whether it be Solar or any other form.

A change of energy from heated water to heated air represents a loss of energy. A change from rotary motion to electric current is also a loss. A change from available energy to stored energy is a loss. Sometimes these losses are minimal and sometimes they are severe, but always they are losses – stones only roll downhill!

To arrive at a logical sequence, and thus to a system with the minimum number of conversions from one energy form to another, it clarifies the picture to see all our energy forms in three stages.

First Stage Represents the incoming energy without conversion.	Sun's heat rays Sun's light rays Air movement Water movement
Second Stage Represents energy with one stage of conversion	Moving hot water Moving hot air Electric current Rotary motion

Third Stage	
Represents one or two conversions, and is usually the point where the energy is taken for use.	Bulk hot water Bulk hot air Bulk hot pebbles Warm earth Hot concrete or brick Charged battery Water stored at a height Methane gas or wood Food

By seeing the three stages set out, the differences become apparent and needless changes of energy form may be avoided. It also follows that Solar Energy should be collected as close as possible to the form in which it is used. And if a storage system is included in your own arrangement, then that also should utilize energy as close as possible to the form in which it was collected. The *least number* of conversions used ensures the maximum energy collected arriving at the point of use.

Consider now the house structure.

Professor Thrombé did so in a most succinct manner when he observed, 'It is a barrier designed to modify the external environment before it reaches its occupants'. He is saying that, whereas we inside always wish to be at 22°C, the outside is always anything but 22°C (72°F). By recognizing this endless battle we can act to achieve superior results from our Solar Energy, or indeed any energy! Fig. 48 shows the general situation with two heat losses and one surprising heat gain.

A beautifully clear exposé of these losses and gains was recently given in a paper by Professor Page of Sheffield University, U.K. (The Professor is the Founder-Chairman of the U.K. Section of the International Solar Energy Society.) To understand it we have to introduce one new technical term, but as it relates to easily recognizable units it will present little problem. It is: that for every degree C difference between the inside and the outside temperature of a building, there will be a loss of so many watts (that's the same watts as are used to rate an electric light bulb) through every square meter of wall or window we are talking about. A typical figure for a single-pane window (closed) is 5.6 watts per meter per degree C difference.

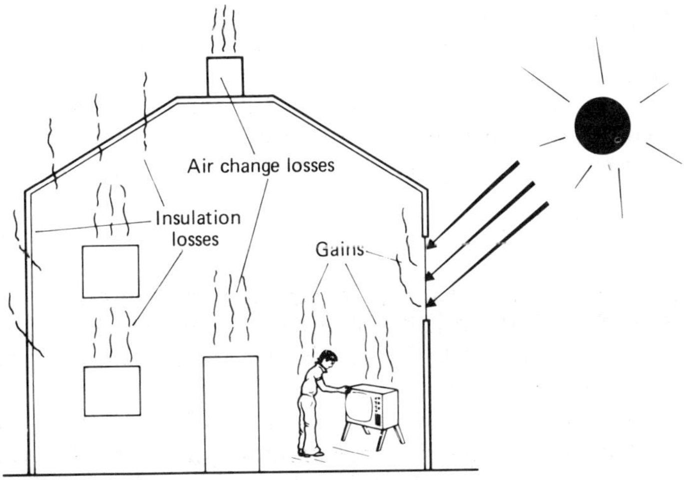

Fig. 48. The house structure: 'a barrier designed to modify the external environment before it reaches the occupants'. Two types of losses and one gain!

This 5.6 watt loss would rise to 16.8 watts per square meter if it were 3 degrees colder outside than inside. Furthermore, if the window is 4 meters in size (which is more likely) and it's still 15°C outside, and 18°C inside, then the loss is up to 67.2 watts. Four times the area has now pushed the loss up four times, so that the loss is now more than the power required to keep a 60-watt bulb burning all the time – just to make up for the heat going out of the window when there is a mere 3°C difference inside to outside.

As a considerable proportion of heat loss data will still be given in British thermal Units, or BtU.'s (a system the British now neither use nor, so they claim, comprehend), an easy conversion is obtained by multiplying the BtU. per square foot per °F (often called 'k') by 6.5 to give watts/meter/°C and thus directly compare with the following.

With the system of measurement explained, we can look at the loss figures for the other parts of the house before seeing the overall picture for the complete house.

Double-glazing halves the normal 5.6 loss to 2.8 watts/meter/°C

Wall in a typical modern house in U.S. is, 1.5 watts/meter°C (which means in some older houses it might be worse)

Wall to some European standards as good as, 1.0 watts/meter/°C

Wall with good insulation added is down to 0.7 watts/meter/°C

Well-insulated roof is also as low as 0.7 watts/meter/°C

A good floor is 0.76 watts/meter/°C

An average mix of a well-insulated wall and double-glazed windows, gives 1.0 watts/meter/°C

From the above, it is possible to work out the individual losses that pertain to your abode, and by adding them all up arrive at a total heat loss for the building due to imperfect insulation. For a well-insulated detached house of 100 meters floor area this is something like 215 watts going out all the time for each degree difference outside to inside temperature. If apportioned over the 100 meter floor area, that's 2.15 watts for every square meter. Comparing this with a poorly insulated house with generous windows, all single glazed, puts the loss up to 570 watts or 5.7 watts per square meter – still for *each* degree *difference.*

All the losses discussed so far relate to imperfect insulation, the seeping out of the heat through the barrier we have erected between ourselves and the external elements. (In summer a similar battle takes place in reverse – then the heat is trying to seep in.)

Now to the second heat loss; the one due to the gradually changing room air necessary to give us adequate ventilation. For all normal situations, it is assumed that the air in a house changes once every hour; in a low-rise house this may come down to half a change an hour, but in a house with vertical chimneys it will get up to two changes an hour. This constitutes a considerable heat loss.

For a house with 2.8 meter high ceilings this loss is taken as 0.93 watts/°C for every square meter of floor space; if the air change is only once an hour; and 1.86 watts/°C if it changes twice an hour. Note that this loss is given in exactly the same way as the 'seeping' loss was finally given – a wattage loss per degree C, and then apportioned for each square meter of floor space. But, of course, this time the losses are the 'carried out' losses as the heat goes out of the door or up the chimney with the changing air.

Now we can see the whole picture of the house heat loss simply by adding the two losses together. Our well-insulated detached house with vertical chimneys has a total heat loss of:

2.15 seepage loss, plus 1.86 carried loss, or 4.00 watts/°C for each square meter of floor space. As you recall this is a 100 square meter house, which means that the total loss going on all the time is 400 watts for every °C/ difference that there is inside to out. Not too bad, when you compare it with the non-insulated house now up to; 5.69 plus 1.86 to total 7.55 watts/ °C – not so good. It should also be self-evident that as the insulation in a house is improved, the ventilation loss gets to be a larger per cent of the total heat loss.

As something to counter the depression that must now be settling over you as you see your warmth disappearing in all directions, there is one actual heat *gain* to cheer about. This comes as a result of paying for energy to achieve objectives other than warming the house. Yes, it still has to be paid for, but one does not usually reckon with the TV, stove, lights, etc., helping to warm the house. An even stranger source of unsuspected warmth is the heat coming from the human body! Did you realize that when you are buying food you are paying mainly for fuel for the body, and that the term 'calorie' is a measure of a quantity of heat just as the watt hour is. It all comes back to heat when you realize that the calories in the food are Solar Energy taken into store by the photosynthesis process (those leafy collectors) as the food was grown. And, there is even a figure for the amount of heat that we put out; on the average, each human body puts out about 75 watts all the time; night and day.

This is a summary of the heat that comes into a house by

any other means than recognizable heaters. These are the average *daily figures* for a 4-person household.

Each body gives out 75 watts, and each spends an average of 16 hours in the house to total	4800	watt hrs.
About 300 watts of lights are on for 6 hours	1800	"
Approximately 160 liters of hot water are used from which half of its heat gets into the house	14000	"
Television, 300 watts on for 6 hours	1800	"
Other electrical appliances (incl. refrigerator) total 250 watts on for 24 hours	6000	"
Cooking stove	7600	"
And finally a Solar contribution; 600 watts a day for each of 20 square meters of glass	12000	"
A total of	48000	watt hrs.

These 48,000 watt hours are the total heat that escapes into the house from these extraneous sources in a day; an average of 2.0 kilowatt (kw.) (2000 watts), all the time. It is this heat that accounts for a 3°C rise in the house that is normally allowed for before working out the size of an intended heat generator.

In detailing the two main heat loss groups and the one heat gain group, a better understanding and through it a better utilization should have been obtained. By working out the loss for each month of the year, an energy budget could be established to compare with the actuals, as measured by the utility billings and the oil consumption.

In summary, the four main areas that contribute to the best use of the available energy are: keeping the number of energy changes to a minimum – prudent selection of heat insulation – care in regulating the number of air changes – and due allowance for the unintentional heat input.

To complete the chapter we are including several important points that we have gleaned from various sources; they are worthy of careful consideration in attempting to obtain even better results from your energy. They appear in barest outline –

mere germs of ideas that may trigger a fertile mind to expand one of them into a practical arrangement.

1. The shape of a building determines the *length* of its outside walls, and thus the heat loss from that source. The walls of a long thin rectangular abode being 30% longer than a round one of identical area.
2. If you can arrange to have some or all of a South facing wall blackened and glass covered (greenhouse effect) then the Solar gain is well worth having.
3. By including thick walls or concrete floors (they can be brick rubble filled), a house's thermal capacity is increased leading to a more stable overall temperature. (Building regulations still have to be adhered to.)
4. The inflow and outflow of heat can be considerably influenced by the use of moveable panels. Example: A large panel hinged at ground level can be swung up at night to cover a large window and retain house heat. During the day it acts as a reflector to bounce more heat in through the window.
5. Arrange the house that any thick brick or concrete wall or floor that receives direct Sunlight is not covered with carpets or curtains; that assumes you wish to collect more heat. The opposite actions should take place if you are trying to keep cool.

CHAPTER 8

The Sun's Contribution

The study of the Sun's radiation as it arrives on earth has occupied scientific minds for years and has filled many weighty volumes. To achieve our goal it is sufficient only to understand how the radiation arrives; to what extent it varies; and what constitutes its main parts.

An examination of Fig. 49 shows that the radiation arrives at a collector by two major and one minor paths. The major paths are the direct rays as felt when the Sun is out, and the diffuse which are present at all times of the day. The minor

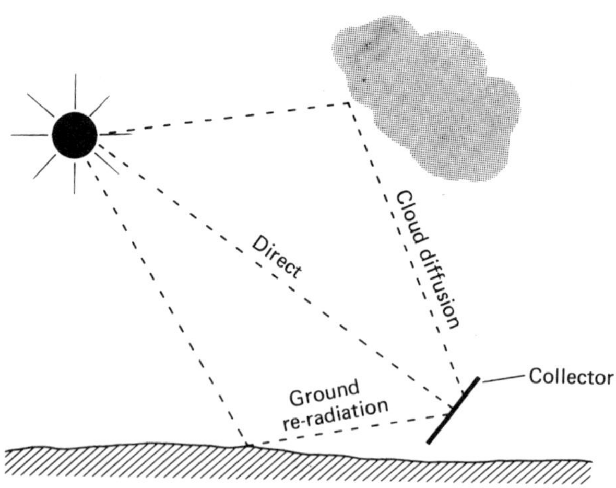

Fig. 49. The three forms of solar energy arriving at a flat plate collector – but the ground re-radiation is generally ignored.

path is a re-radiated energy that emerges from the ground towards the end of the day as a result of the day's Sun. (We really should be more grateful to these radiations than we are, as they are vital in preventing the earth's night temperatures from plummeting to an untenable depth.) But, in most Solar Energy work only the direct and diffuse parts are considered – and we are no exception.

As the diffuse radiation is a reasonably constant component, the relative strength between it and the direct varies considerably. When the Sun is out, the direct rays have about six times the strength of the diffuse: when it is in they have only half the strength.

Consider now the direct radiation that arrives in the winter as compared with the summer. See Fig. 50. This shows that the low winter Sunshine rays on the left are considered to be the same strength as the rays at the end of a summer day. That sounds like an extraordinary statement, yet is quite true. Winter is not cold because the Sun is further away and the rays are weaker. Winter is cold because the air and the ground are cold. The important point is that whenever the Sun is at a particular height above the horizon, the radiation that arrives on Earth is the same. By understanding how that energy arrives, the reason is clear; our Fig. 50 shows it pictorially.

At all times the energy arriving at the *commencement* of our atmosphere is close to 1400 watts per square meter. (That's the same measure we used to show the *rate* of heat loss from a house – but, as you see, this is coming in at a whopping rate.) Unfortunately, in passing through the atmosphere it is reduced to 1000 watts/sq. meter, but this is still a very considerable energy rate. After one hour at this rate 1000 watts, or 1.0 kw/hr, or one unit of electricity (equivalent) has arrived – on just one square meter!

This 1.0 kw/sq. meter is the total energy, direct and diffuse, that will arrive at that square meter if it is facing directly at the Sun, *and* if the Sun is fairly well overhead. When the Sun drops down nearer the horizon, the level drops down (to 700 watts in our example) because the rays now have to pass through the

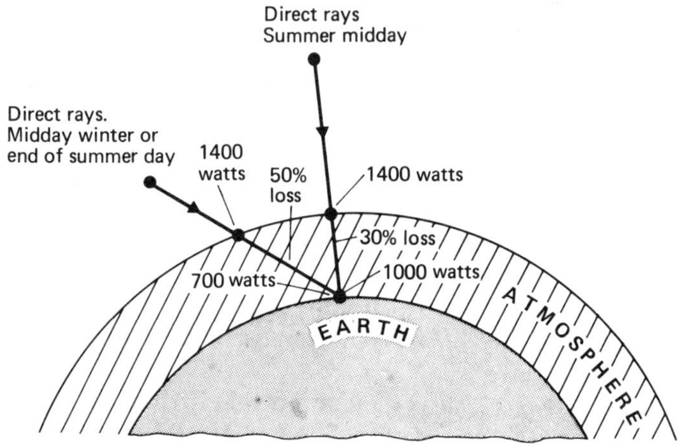

Fig. 50. Summer afternoon sunshine is no stronger than midday winter sunshine.

atmosphere on the slant, and thus have a *longer path* to travel through the attenuating atmosphere. It is the longer path that gives weak Sunshine when the Sun is *low*.

To achieve the 700 watts/sq. meter still requires that the collector faces the Sun, which means it has to dip down towards the horizon and move around to the West. In practice this extra complication of 'tracking' the Sun is rarely carried out, although the loss of watts is considerable.

As well as considering the paths that the Sunlight arrives by, and the strength it arrives at, the difference between 'light Solar Energy' and 'heat Solar Energy' warrant discussion. Imagine that the Sun's rays are composed of many similar rays and yet all are slightly different; ranging from infra-red to ultra-violet. If this range is likened to that covered by the eight notes in one octave of a piano, we can then refer to the rays as if they were the eight notes and see them in more familiar terms. Fig. 51 shows the comparison, but at first sight, ignore the part to the right.

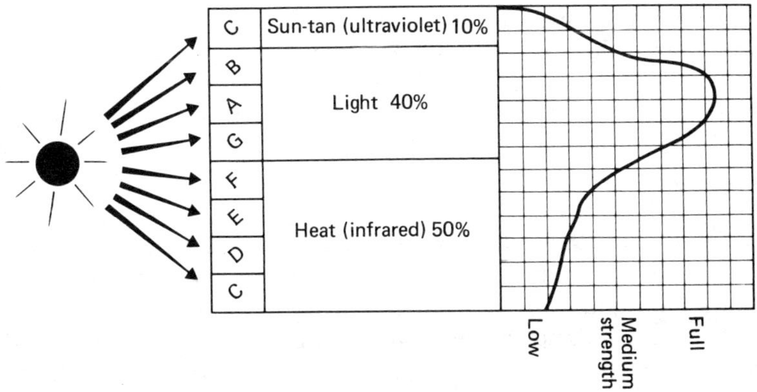

Fig. 51. Full sunlight splits up into three main groups, and they are all of differing strengths.

As you see, full Sunlight is akin to all eight notes being played at once – a sufficiently excruciating thought to mark the dullest memory. The rays GAB are the light rays, and CDEF the heat. By rotating the paper anti-clockwise, it can be seen that most of the light rays are not only quite separate from the others, but have their own identities and strengths. (A re-cap of the famed greenhouse effect makes further sense now. Observe that as glass only allows through rays EFGAB, and it is rays CD that are bouncing back, then they cannot get out.)

It is becoming clearer now, that cloud cover could (and does) shut out heat rays CDEF, and yet still allow light rays GAB to pass through and operate electricity producing Solar cells.

So far we have seen that there are two main routes by which energy arrives at a collector, and that one of these (direct) can be split into three groups; heat, light and U/V. We conclude the chapter by defining the total value, or strength, of the Solar Energy as it arrives on earth, before we go to the more practical chapters.

Earlier, we saw that at all times the energy arriving at the start of our atmosphere is 1400 watts/meter and if the Sun was high overhead we would receive 1000 watts onto every square meter we held out. It may have become apparent, also, that as we could not run round the earth to keep our square meter under the Sun, there would be a falling off of watts at each end of the day, the energy finally reducing to zero as the Sun dropped below the horizon. Therefore, to put that data into a useable form, we use a method that totals up all the day's watts and presents it as one figure. This figure represents 'the total energy received onto a flat plate collector in a day, in kilowatt hours per square meter'. Note that we have moved from kw/sq. meter to, kw/hr/sq. meter. The former is a rate of *flow* and the latter, by having a time added to it, is an *amount* of energy. (Just as when a tap is flowing *fast,* there is no indication of the amount of water flowing from the tap, until a *time* it was running for was added.) The amount of energy collected in kilowatt hours (kw/hr) during a day conveniently ties in with the previous terminology used to examine, heat used and heat lost.

The information from Fig. 52 onwards is based on actual readings taken at measuring stations, their recording devices totalling up the day's energy received from all sources. While the stations measure every day, it is the average *daily* figure for a particular month that is shown. This gives a sufficient guide to the gradual change over the year at that particular location. Note the daily average for the year at the right of each graph. By multiplying this by 365, the total energy received in a year is obtained. This will come out to between 1000 and 2200 kw/hr per square meter, per year.

While in Fig. 52, we do give the best angle for a collector in different latitudes, there is little doubt that an existing roof angle (or wall in the higher latitudes) makes a very strong argument in favor of its angle. Give strong consideration to using what you have, but increasing the area of collection.

The map of Fig. 52 shows the locations we have selected to provide a detailed month-by-month energy budget. By using this first map in conjunction with a similar map in Chapter 17 (showing the Solar stength over the whole country) you can select the most appropriate detailed energy budget for your

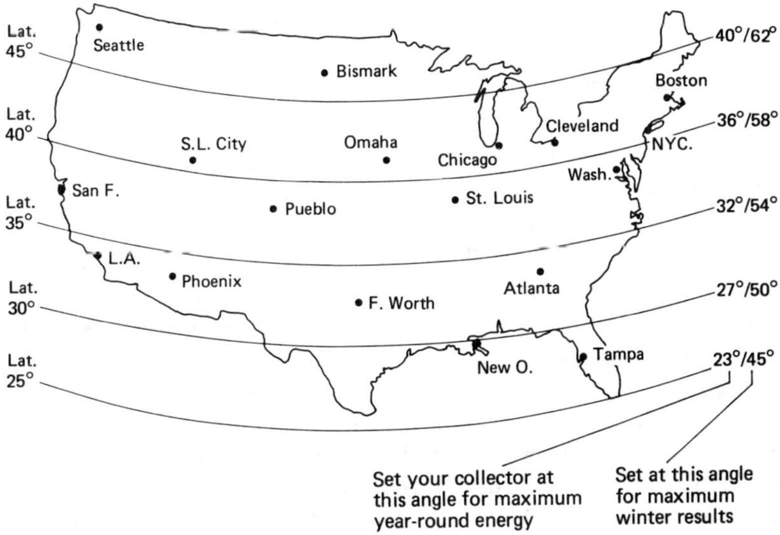

Fig. 52. The following graphs give a yearly energy budget of incoming solar for each of the above U.S. towns in alphabetical order, followed by others around the world.

area if you are not close to a selected town. If you can obtain local data this is, of course, better still.

When you have located your most relevant data page, you will see that it shows the energy you can expect from a horizontal collector for each month of the year. The necessity of looking at the energy arriving in each month is now apparent. Total energy over the year may not be the best criteria, it is energy when you need it that matters. Do not forget that the figure shown is the energy that arrives at that point. Most likely you will only be able to grasp about 50% (fair) to 70% (very good) of it for home heating.

If you find data that is more local than we have supplied, and wish to convert it from the various forms that we encountered, this is how to do it. But, be careful that the conditions they were measured under were identical; angle of collector, etc.

They invariably only give energy that arrives on a *horizontal* surface.

If the data given in were:

miliwatt hrs. per sq. centimeter, divide by	100	= kw/hr/sq. mtr.
calories per sq. centimeter divide by	85	= kw/hr/sq. mtr.
langleys divide by	85	= kw/hr/sq. mtr.
BtU. per sq. foot divide by	313	= kw/hr/sq. mtr.
Megajoules per sq. meter divide by	3.6	= kw/hr/sq. mtr.

Be particularly careful that the figure you convert is the total energy *amount* that arrives in a day and not the *rate* of arrival.

There is one point on the Earth that receives more Solar Energy in a month than any other point and far outstrips any other location in our listings, but the chances of you guessing it, or indeed being interested in it are nil – it's the South Pole! Well-documented records show that all through the month of December, it has 10.0 kw/hr of heat per day arriving on every square meter. What has happened, is that the Earth has rocked to the extreme of its tilt relative to the Sun, so that it is shining on the pole 24 hours a day; that plus the clear skies does the rest. It is only cold down there because the air is cold. Solar rays go straight through air without warming it, but they do warm the earth, it is the warm earth, and sea, that warms the air.

This artic surprise underlines the fact that you should try to stop judging the amount of Solar Energy by the outside *temperatures.*

Looking at Fig. 73, you will see that although the U.K. has a poor reputation for Sunshine (it gets about 4 hours on an average every day of the year, 1450 hours a year), it does receive a useful average of 2.8 kw/hr per meter for each of its 365 days.

In a year that one meter will have received over 1000 kw/hr, an amount that is delightfully compared in one textbook with 'the equivalent work carried out by one pair of bullocks in a year' – presumably, this discounts the benefits of fertilization bestowed by the bullocks, but is otherwise comparable.

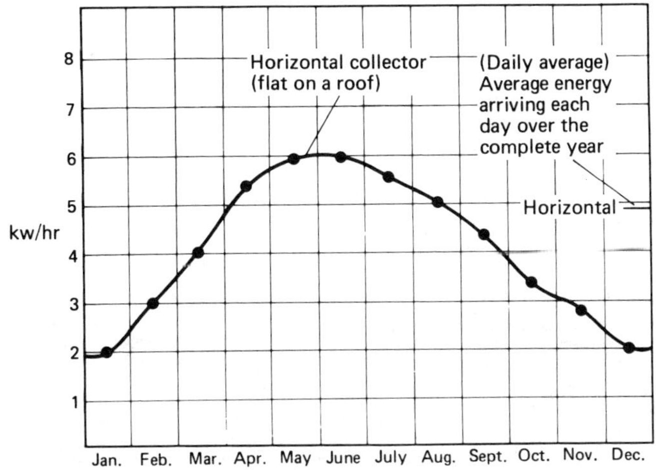

The point marked in the center of each month is the average energy that arrives every day in that month

Fig. 53. U.S. Atlanta. Lat. 33°. (The reading from the graph is the amount that arrives each day onto a square meter.)

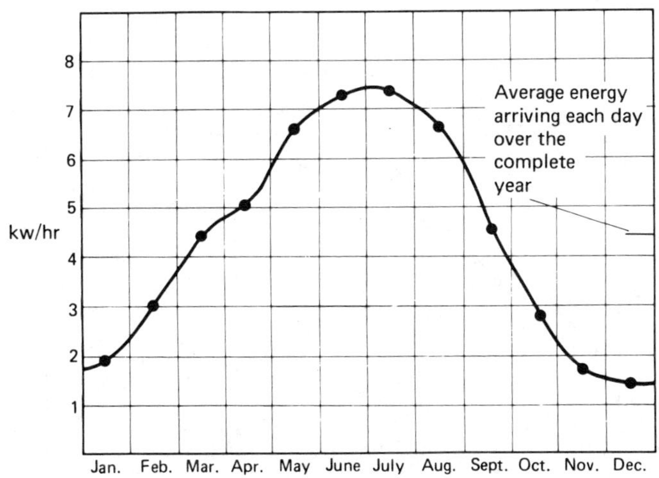

Fig. 54. U.S. Bismark, Lat. 47°.

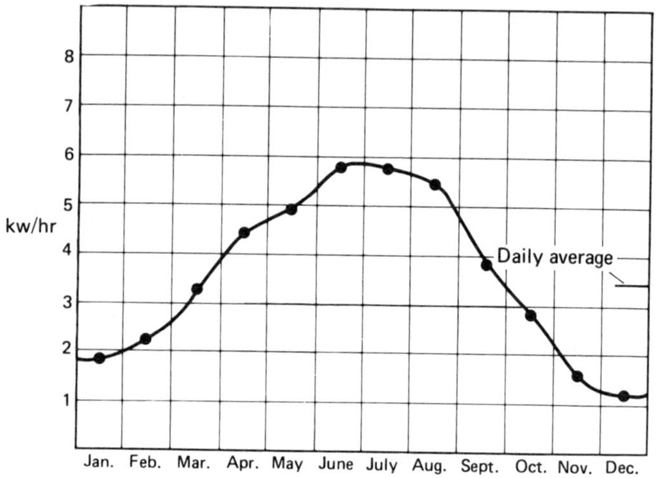

Fig. 55. U.S. Boston (Blue Hill). Lat. 42°.

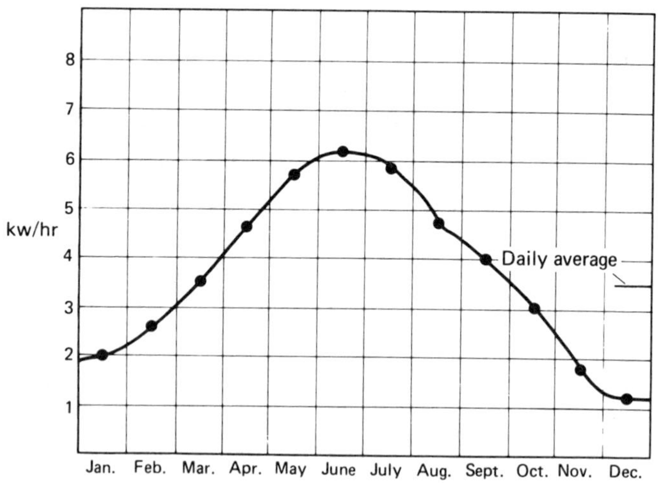

Fig. 56. U.S. Indianapolis, Lat. 40°.

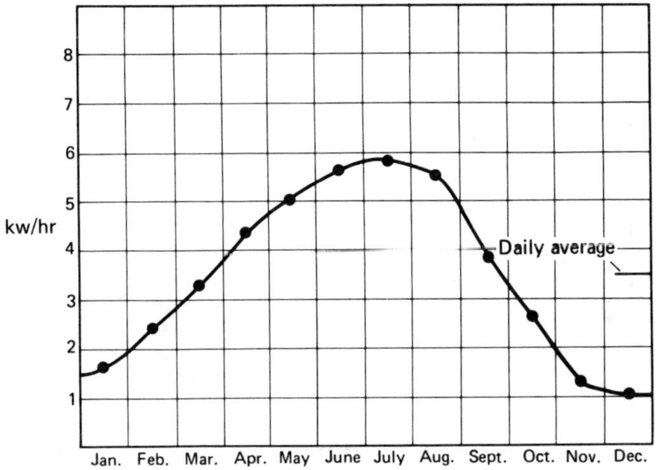

Fig. 57. U.S. Cleveland, Lat. 41°.

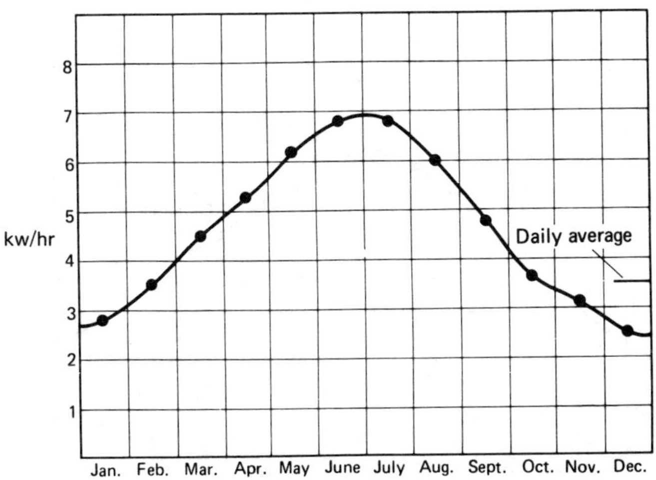

Fig. 58. U.S. Fort Worth, Lat. 32°.

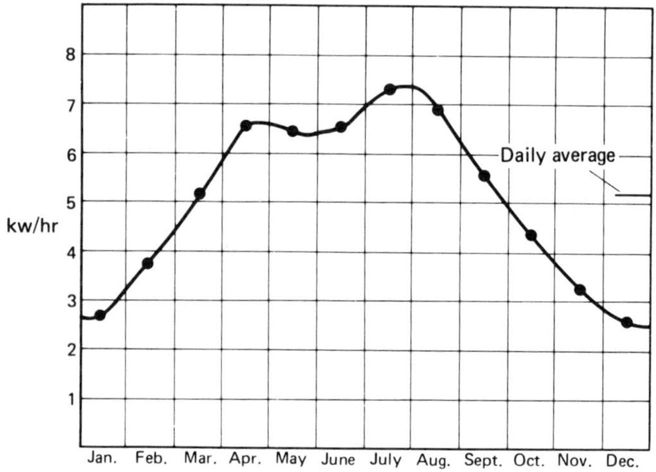

Fig. 59. U.S. Los Angeles, Lat. 34°.

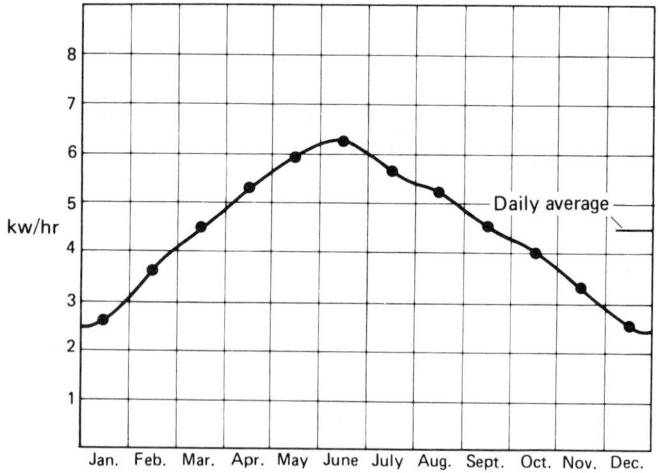

Fig. 60. U.S. New Orleans (Lake Charles), Lat. 30°.

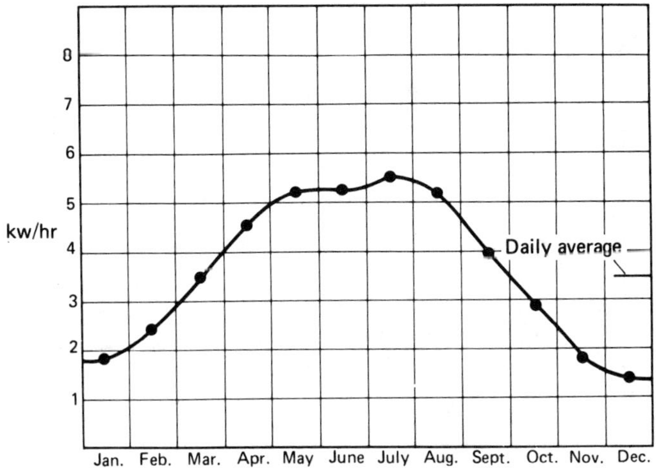

Fig. 61. U.S. New York, Lat. 41°.

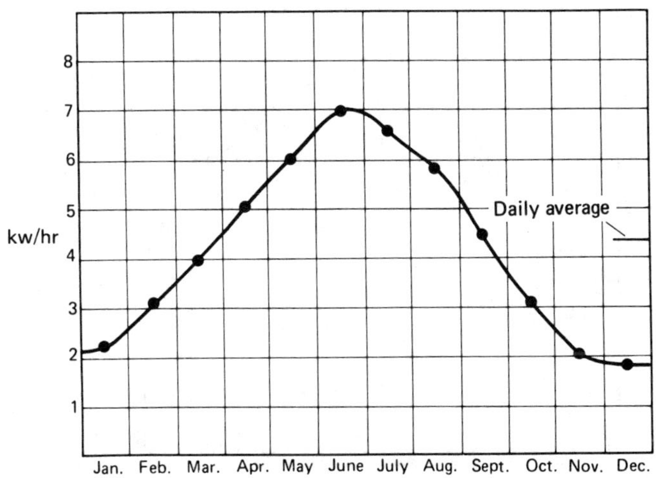

Fig. 62. U.S. North Omaha, Lat. 41°.

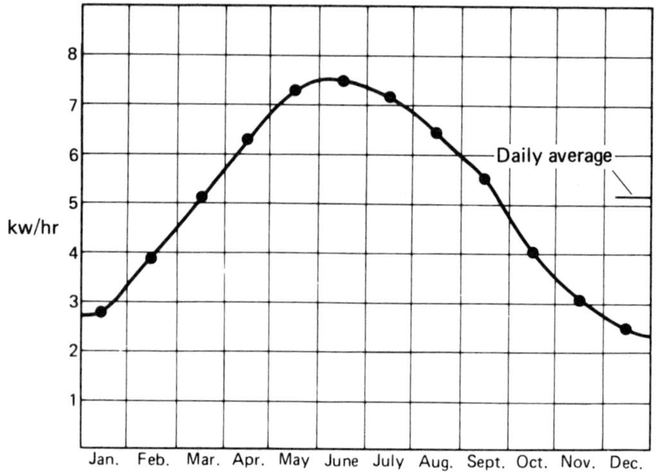

Fig. 63. U.S. Pueblo (Grand Junction), Lat. 39°.

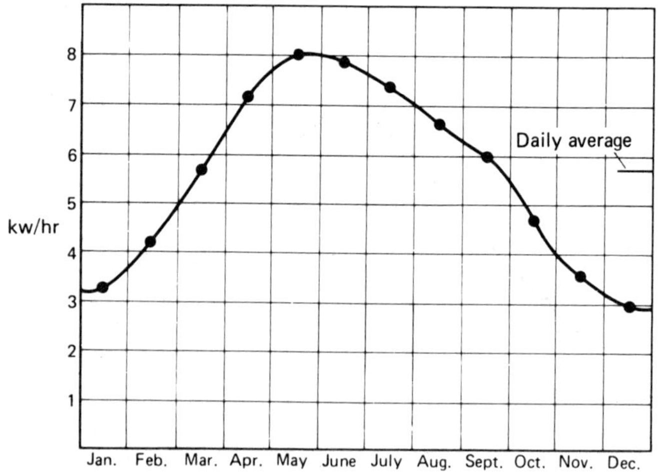

Fig. 64. U.S. Phoenix, Lat. 33°.

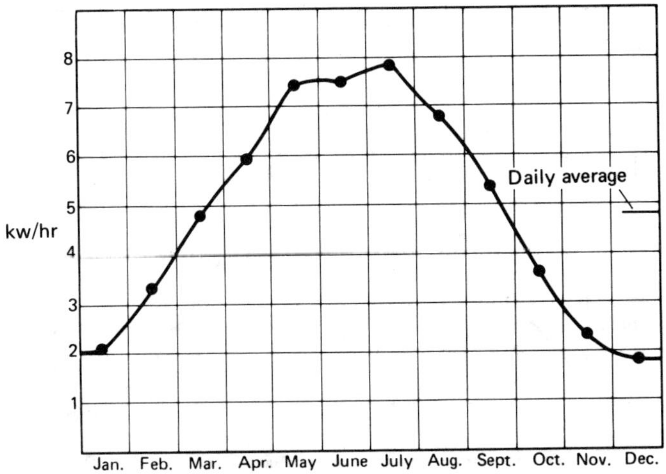

Fig. 65. U.S. Salt Lake City, Lat. 41°.

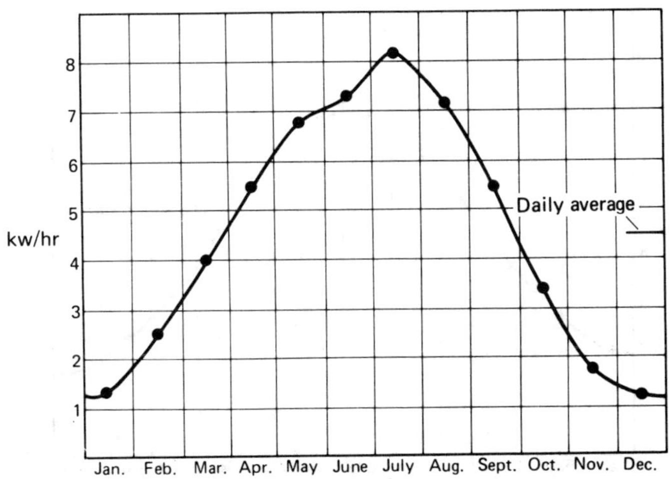

Fig. 66. U.S. San Francisco Area (Medford), Lat. 47°.

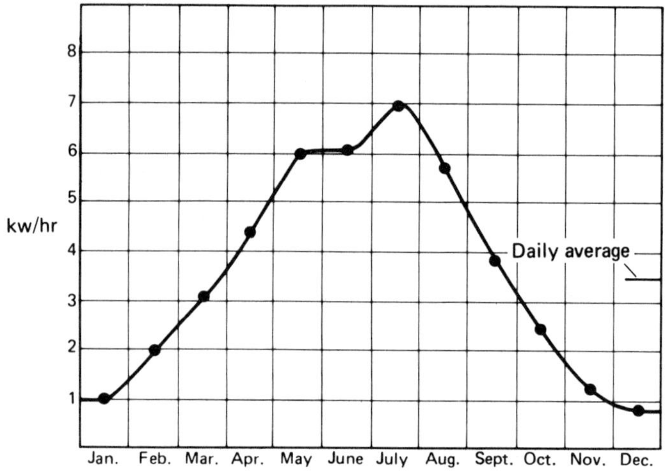

Fig. 67. U.S. Seattle (Seattle Tacoma), Lat. 47°.

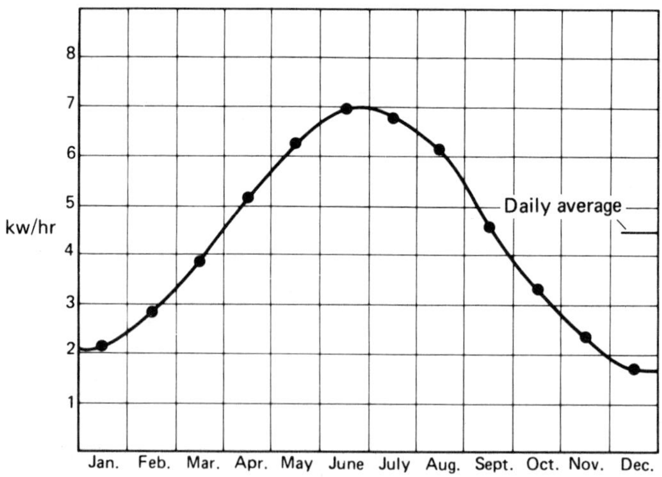

Fig. 68. U.S. St. Louis (Columbia), Lat. 39°.

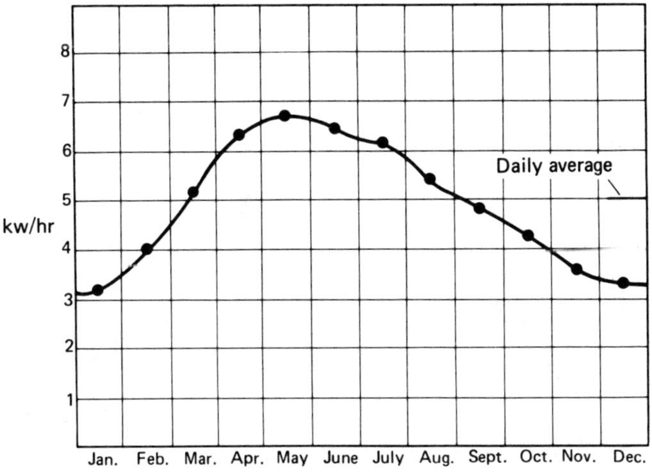

Fig. 69. U.S. Tampa, Lat. 28°.

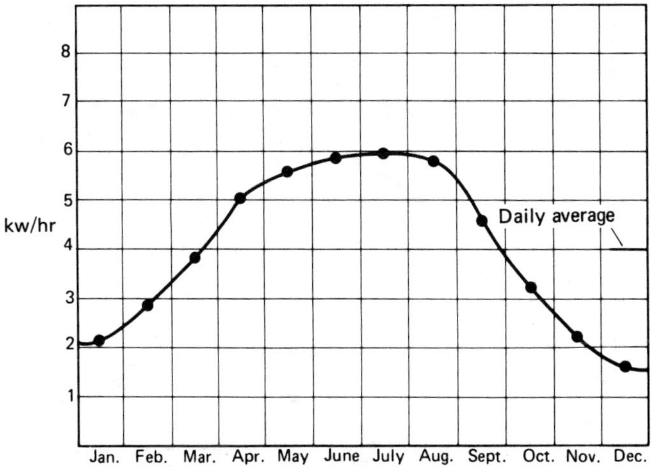

Fig. 70. U.S. Washington D.C. (Sterling), Lat. 38°.

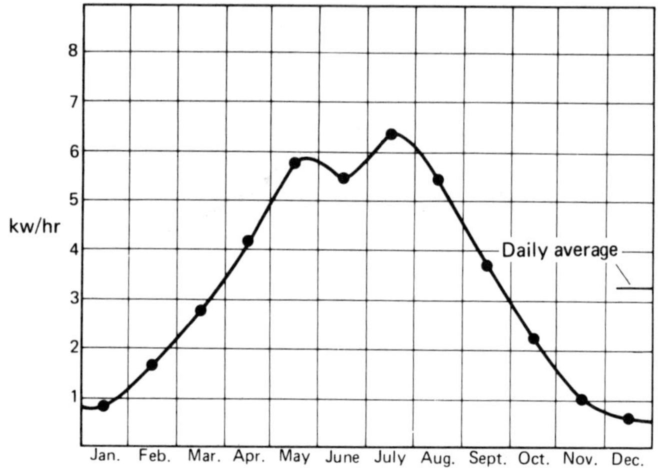

Fig. 71. Canada, Vancouver, Lat. 49°.

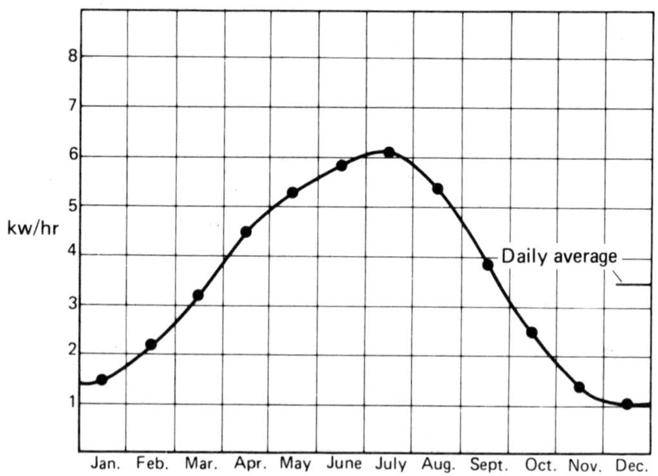

Fig. 72. Canada, Toronto, Lat. 43°.

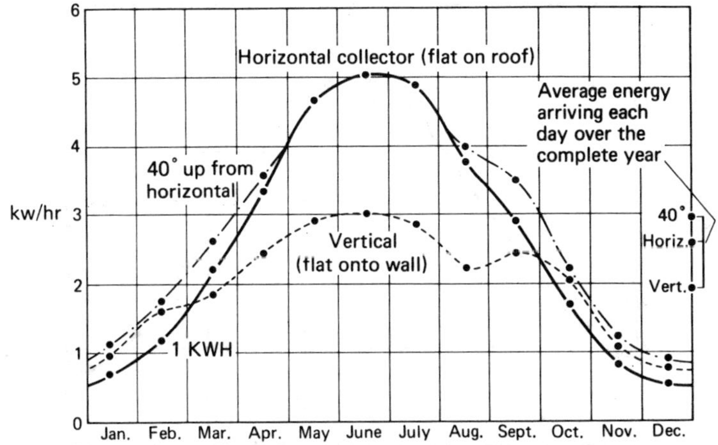

Fig. 73. U.K. Cambridge, Lat. 52°.

The main line in this graph compares with the line in all the other graphs and is the energy measured on a horizontal collector. The other lines show energy on a vertical collector and energy on a 'best angled collector'. If your collector is not horizontal a 'guesstimate' correction applied to your towns graph will give more accurate information.

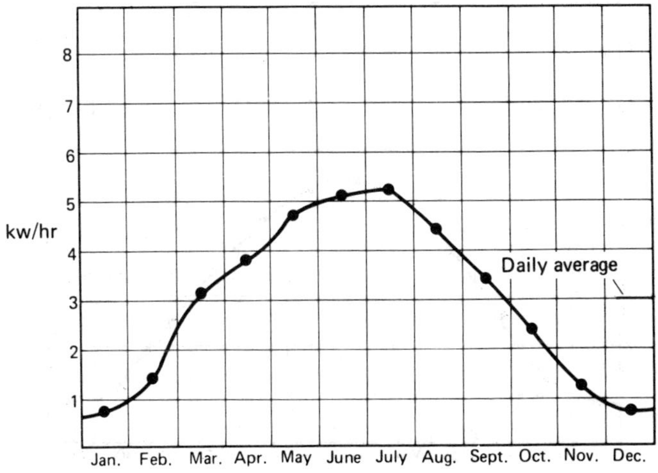

Fig. 74. France, Paris, Lat. 48°.

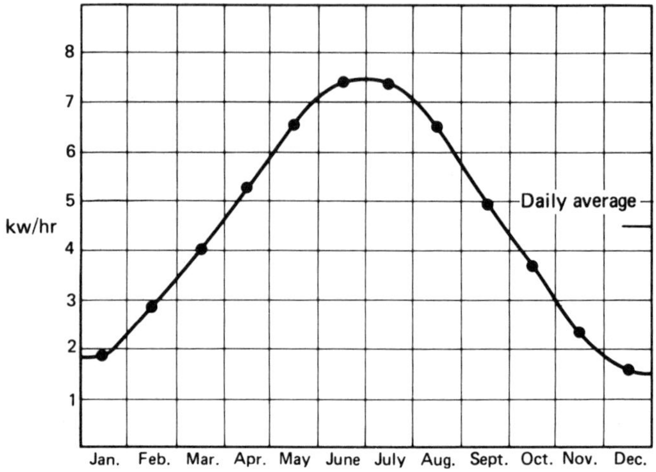

Fig. 75. Italy, Rome, Lat. 42°.

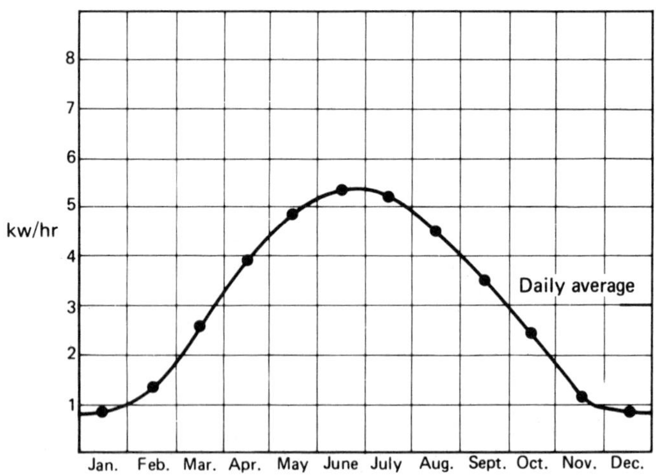

Fig. 76. Germany, Wurzburg, Lat. 49°.

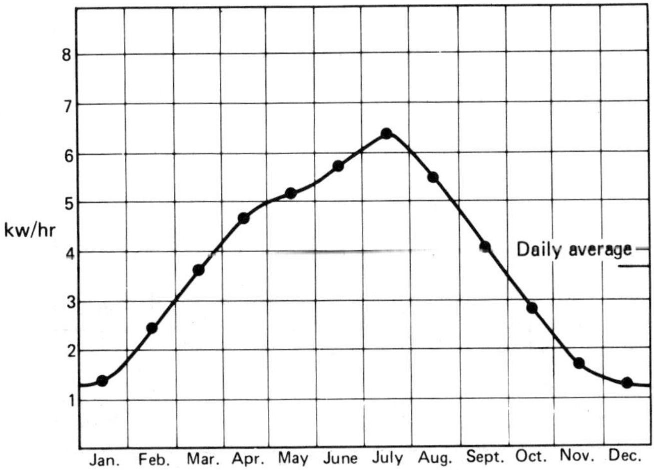

Fig. 77. Switzerland, Zurich (Locarno-Monti), Lat. 46°.

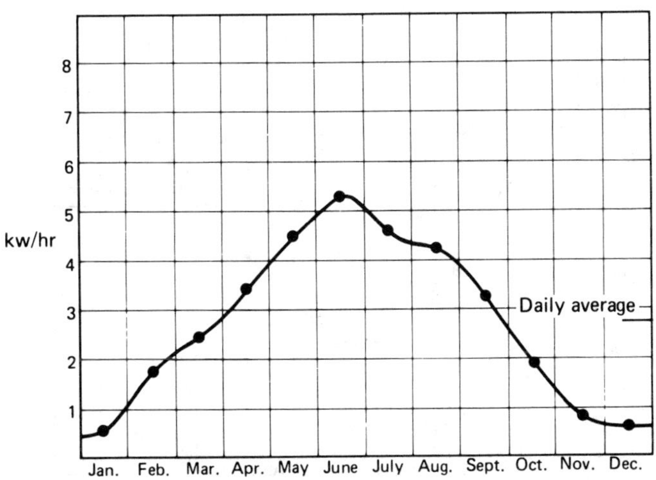

Fig. 78. Netherlands, Amsterdam (De Bilt), Lat. 52°.

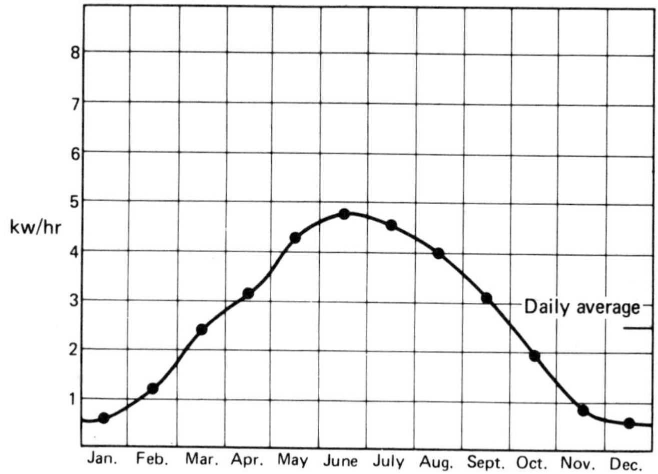

Fig. 79. Belgium, Brussels (Uccle), Lat. 51°.

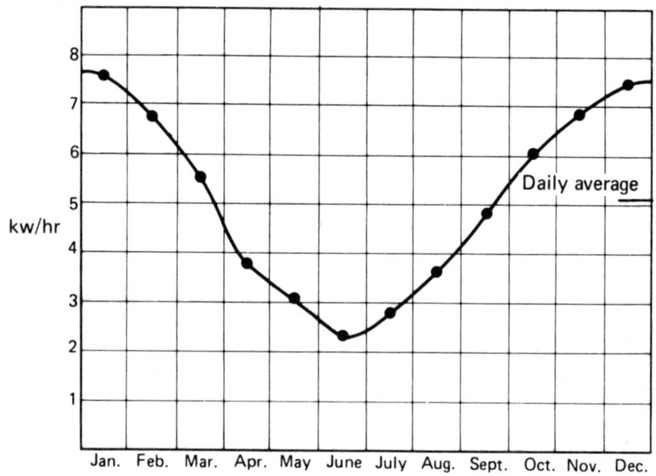

Fig. 80. Australia, Perth, Lat. 31°.

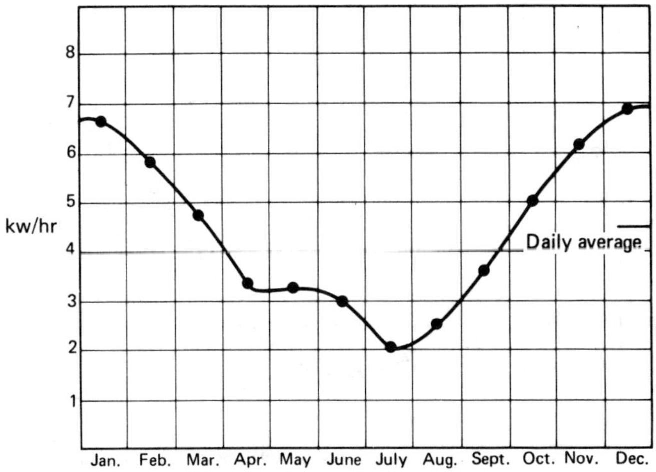

Fig. 81. Australia, Melbourne, Lat. 38°.

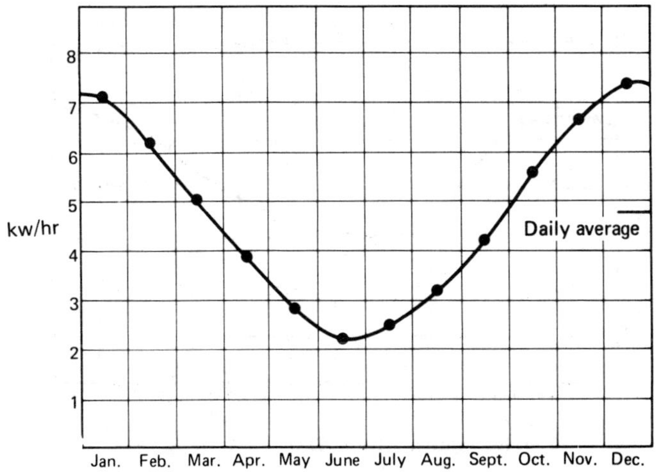

Fig. 82. Australia, Griffith, Lat. 34°.

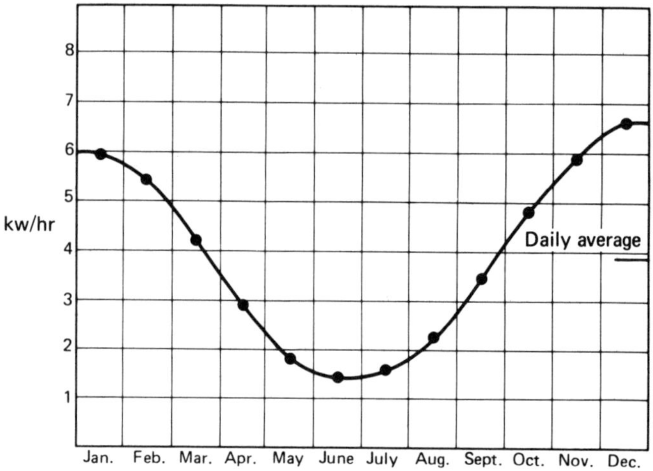

Fig. 83. New Zealand, Wellington, Lat. 41°.

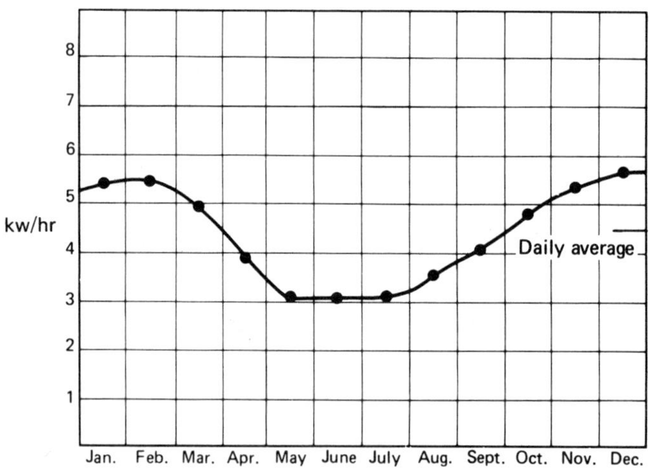

Fig. 84. South Africa, Durban, Lat. 30°.

SEE IMPORTANT NOTE UNDER FIG. 73

CHAPTER 9

Producing a Solar Collector

To implement our aim and spawn home Solar systems requires that we cater to three groups: do it yourself; buy in components to assemble yourself; or have a complete system fitted for you.

To accommodate the first group we will examine the detailed construction of some imaginary collectors; an approach that encompasses many more possibilities than if we had restricted ourselves to two specific designs, thus excluding all but those people who were exactly suited. For those who are interested, the Directory leads to information on plans that are detailed down to the last nut and bolt.

For our second group, the Directory leads to ready-made collectors that can either be purchased separately, or as part of a complete hot-water and home-heating kit.

The third group can also obtain guidences from the Directory section.

Now we move from the theoretical to the more practical chapters; a division that is not too abrupt as we still include the reasoning behind each choice of method or material. The remainder of this chapter guides you through the construction of your own collector unit, a unit made from readily available materials which you can fashion into a form that suits your precise needs.

With a dozen different parameters to select from, each having two or more options, the varieties of collector designs that could be produced are endless.

To reduce the options to manageable proportions, and at the

same time guide the reader to the most successful designs, requires that some decisions are made for you. The option chosen, and the reason for choosing same, are both given.

Flat Plate or Concentrating Collector? Flat plate chosen.

Concentrating collectors have so little following that this is our first reference to them. (If only to show that we didn't forget; one form is a 1-meter diameter, saucer-shaped mirror facing the sun, which at its focus point about half a meter away, produces a small area of concentrated heat.)

Open Plate or **Greenhouse?** (glass covered) **Greenhouse chosen**

Only in rare circumstances can the losses associated with an open collector be tolerated. The glass cover not only traps the heat but also helps to reduce heat losses caused by cross winds.

Liquid or **Air-filled? Both used**

Few air-filled collectors are used, but they are gaining, while they are a little thicker, the trunking to them is much more difficult to accommodate than water pipes. Conversely they don't leak or freeze – are much less liable to corrosion, and are lighter!

Integral Storage? No storage in the collector

As some form of storage is included in nearly every system, one option is to include that storage as part of the collector unit. But for temperate climates this is the opposite of what is required, as it makes it difficult to take in small amounts of heat between cloud breaks, without losing heat during cloud cover.

Storage in the collector for hot climate operation is quite acceptable.

Single or Double Glazing? Single chosen

The additional cost and complication of double glazing a collector is doubtful as the extra sheet of glass, while retaining more energy, also excludes more of the incoming energy. At collector temperatures below 50°C there is nothing to be gained: it has also been established that double glazing and a selective surface are worse than single and a selective surface. Only if your area is cold and windy, and you wish to run above 50°C should double glazing be used. Incidentally, an idea worthy of consideration is to make the inner sheet from clear plastic.

Open or Closed Waterways? Closed chosen

Making a collector as a complete unit pre-determines that the closed method is used. Closed means that no water can be seen through the glass cover. Open waterways are also much larger assemblies built *in situ* to form an integrated part of a complete roof (described later).

Purchased or Self Constructed Plate? Both discussed

The plate, which is the heart of the collector unit, can be purchased as a finished product.

To summarize what we have said: the collector we are going for is: flat plate, single glass covered, water operated, minimal water storage, covered waterways and with a purchased or built plate. Our mythical collector is something of a compromise as the single glazing is more suited to hot climates (bands 1 and 2), and minimal water storage is more suited to temperate climates (band 3).

Later in this chapter, we go through the construction of your own plate, but now we discuss the several points that will help you in the purchase of a collector plate. They are:

Materials used. Copper, painted steel, aluminum, galvanized iron or stainless steel are all possibles, and should give long life without corrosion (a vital aim) if the correct precautions are taken. Be wary that any additives put in the system water are suitable for the chosen metal and any other metals in the system. Be careful when considering the finish on the metal, ordinary paint is acceptable if it is the way of achieving the black color. IT is not acceptable as a protective finish on plain steel. The new acrylic paints and PVC finishes are much better (for aluminium or steel), but the best is paint that uses silicone resin as the binder. As a further guard against corrosion, obtain clear assurances from your would-be supplier on the following. Ask if it is still acceptable to use his particular collector plate after having described your system and particularly highlighting the different metals with which the water will come into contact. Also remember, it is the finish on the unseen inside that is most important. That's where we are really talking about when we quiz the manufacturer about the expected corrosion-free life.

Complexity of waterways? The more complex these are the more heat the plate will pick up – thus giving a greater per cent efficiency. But greater complexity often means greater resistance to the water flow, necessitating a larger pump to push the water through. The flow resistance is shown as the pressure drop across the plate from input to output. The higher this is, the larger the pump you require. The efficiency of the plate should be given as a per cent (the higher the better), but be careful to note at what temperature this is given, as all collectors get more efficient at lower temperatures.

Water capacity? Professor Heywood noted the following in his very first experiments way back in 1947. In temperate climates where sunshine comes in spasmodic rather than long unbroken spells, too many liters of water actually in the collector plate give an excessive time lag before transferring the heat into store. The effect being to lower the efficiency during intermittent cloud conditions, although the plate might be very efficient in steady sunlight. Again, no problem in hot climates but a vital characteristic if one is to get the most out of a collector in temperate climates.

Black finish. The metal that the plate is constructed of has to have a black finish on the Sun side to absorb maximum sunlight. It is preferable to leave the back bright, as it then radiates less heat. Four possible methods are considered to get the black finish. One, *aluminum* (only) can be black anodized by the manufacturer, and although it often fades to a greyish color after a year or two, it still appears to give quite good results. Two, a *black paint* can be baked onto any of the metals and should stay on the longest. Three, *matt black paint* can be brushed or sprayed on and will give results that vary with the type of paint used. Silicone resin paints are the best as a long-term answer. Four, a *selective black finish* applied by the plate manufacturer. This should stay on well and has this unique feature.

A selective black finish improves the heat gathering efficiency of a plate by taking in nearly as many heat rays as ordinary black paint, but *re-radiating back much less* to the glass cover. A reasonable performance would be a plate that takes in 85% (absorbance) and reflects back 15% (emittance). In case this new characteristic confuses you, we should help you recall that while the glass was supposed to stop all this reflected, or re-radiated heat, it did not stop it *all.* Reducing the amount being reflected back helps the glass do one of its jobs.

To summarize: A good collector plate for use in our collector unit is: long-term corrosion free, has efficient waterways, holds the smallest amount of water (temperate climates) and is preferred with a selective black finish.

For those who wish to go further back to basics, there now follows the detail of how to construct your own collector plate. We talk in general terms but include sufficient practical information to enable you to make your own plate. Even those purchasing a plate to build into a complete collector unit will find the discussion valuable as a purchasing aid. An axiom of the famed Macy's purchasing department is; aim to know as much about your supplier's product as he does before you attempt to buy it!

In the construction of the collector plate that follows, and in the complete collector which follows that, one point should be kept in mind above all others. Collectors must stand up to all weather conditions for years to come. That bears repeating. Collectors must stand up to all weather conditions for years to come.

Making a collector plate. The most likely materials for the construction of a plate are galvanized iron in *corrugated sheet form;* or copper, or galv. iron, and possibly tin plate in *flat form.* These give us two distinct designs to explore, although both are still referred to as flat plate collectors to differentiate from the curved concentrating collectors.

Corrugated sheet designs. The object of a collector plate is to present itself to the Sun's rays to absorb maximum heat, and then have that heat carried away by some medium running behind it (air or water generally). It follows, therefore, that as the Sun heats all the plate, as much of the plate as possible should have water behind it. Fig. 85 shows three methods of achieving this.

(a) Only suitable for hot climates as the quantity of water between the plates is considerable.

(b) Acceptable for all climates, but flat galvanized sheet may be difficult to obtain, and a copper sheet in its place is not recommended.

(c) The best for temperate climates as it has the lowest water capacity; also suitable for hot climates.

Construction Notes

1. Use the natural size of the sheet which will often give a plate of nearly one meter (3 ft.) wide and two meters (6½ ft.) long; as large as one should go.
2. The line of rivets in each valley must be soldered over at top and

bottom to prevent leaks.

3. The 2-meter long open sides have to be soldered to form a sealed finish. Version (b) can be clamped together and soldered, but (c) should have a 'U'-shaped channel soldered down each side.

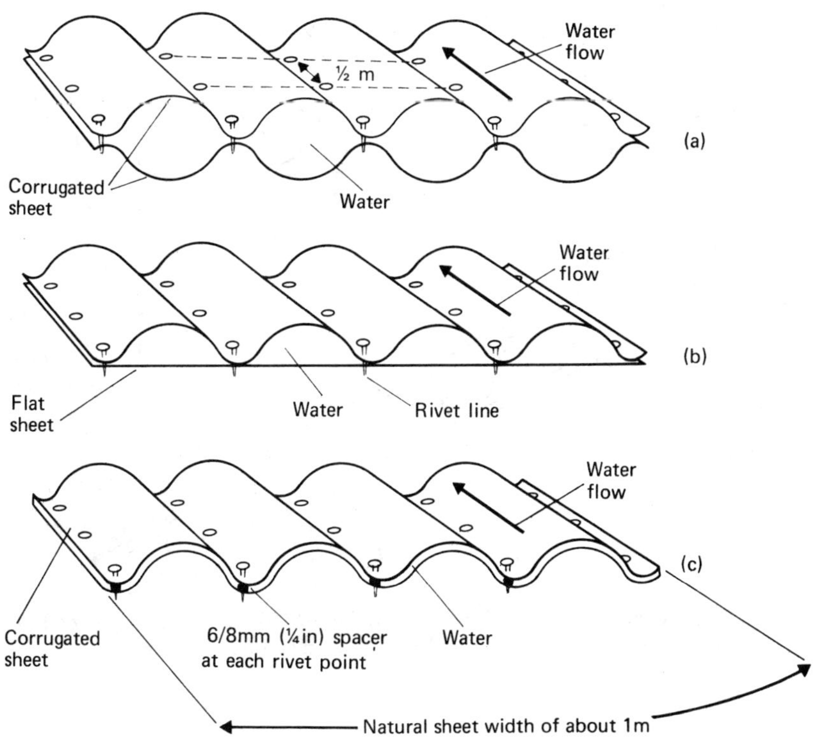

Fig. 85. Three methods of making a collector from corrugated sheet.

This finishes the construction of the center part. It now remains to add some means to feed the water in at either end, so as to complete the whole plate. As water has to enter *each* of the corrugations between the rivet lines, a header box is added at either end of the 2-meter length. Fig. 86 shows the arrangement plus all important dimensions.

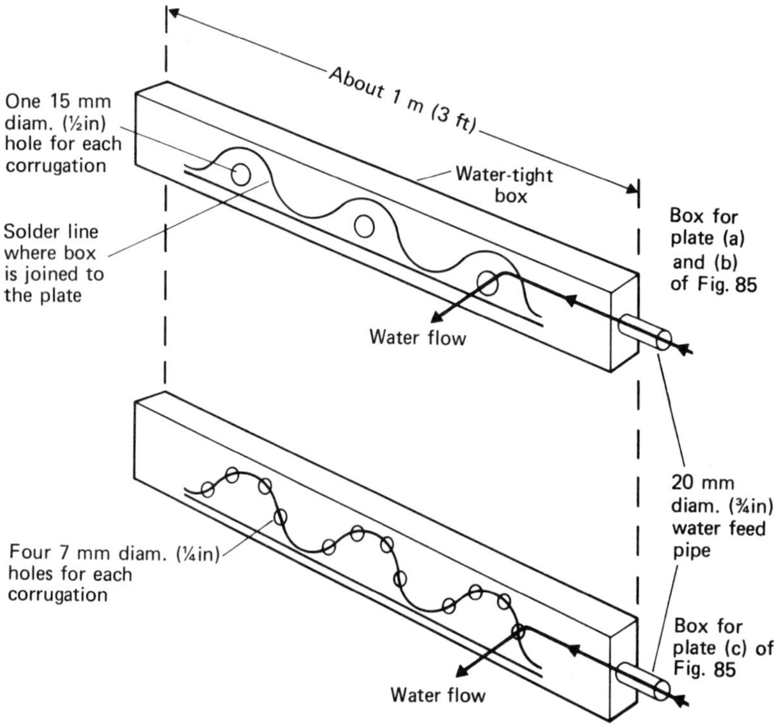

Fig. 86. Header box assembly, two to be added to one plate assembly. Keep box size to minimum for collector use in temperate climates only.

Notes

1. It is advisable to include a gauze filter at each of the 20 mm. feed pipes to strain out dirt. Once dirt gets into the headerbox it is not easy to remove.
2. If the 20 mm. (¾ in.) pipes are reduced in size, the gauze filters should be left in an oversize portion of pipe (i.e. still at 20 mm.); otherwise; they will reduce the pipe bore further still, and impede the flow.
3. The header boxes are generally fitted so that the two 20 mm. pipes emerge at diagonally opposite corners of the plate. This reduces piping when connecting several plates together.

4. After fitting the two header boxes, the complete plate should be checked for leaks by filling it with water.
5. Paint with black on one side only.

This completes the group of plates based on corrugated sheet, and we now move on to consider collector plates built of flat sheets of metal.

Because it has the best heat-conducting properties, sheet copper is an ideal material — but it is expensive. Galvanized iron or tinned plate, if it can be obtained, is quite acceptable as a second best. This group of plates operates by facing the flat sheet to the Sun and allowing the water-filled tubes, bonded to the back of the plate, to take away the heat in the usual manner. Fig. 87 (Parts A and B) shows the two most favored arrangements.

Notes

1. If the spacing between adjacent rows of the 10/15 mm. pipe gets greater than 100 mm. they will not remove sufficient heat from the flat plate.
2. If the bends of the 10/15 mm. pipe are tighter than 100 mm. diameter, dirt traps and other flow restrictions may be created.
3. The pipes in (A) and (B) are normally soldered to the flat plate along their full length and on both sides. The solder is both a heat conductor and a mechanical support.
4. The pipes can be strapped to the flat plate by hooped clips, but it is still desirable to bond the entire length of the pipe to the plate with a heat conducting compound. Plain epoxy resin will not suffice, but heat transfer versions are available.

Two new ideas have recently been devised to increase the effectiveness of flat plate collectors by additions placed between the glass and the black metal plate.

In the first arrangement a glass or clear plastic honeycomb fitment is used to nearly fill the available space. Each section of honeycomb would typically be 5 mm. by 40 mm. with the larger dimension running East to West across the collector face. For an air space between the glass and plate of 25 mm. one

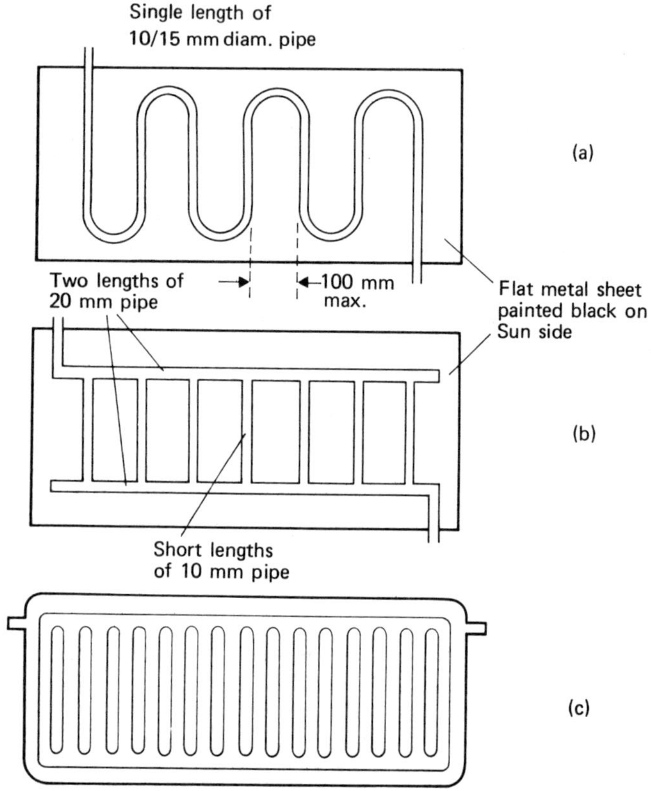

Fig. 87. Three examples of a plate which forms the heart of the complete collector unit; (a) and (b) are made from flat sheets of metal; (c) is an ordinary pressed sheet domestic radiator. They make quite useful collector plates.

would fit a 21 mm. thick honeycomb with a 1 to 2 mm. gap between it and the black plate, leaving a 2 to 3 mm. gap between the honeycomb and glass.

The object of the device is twofold. One is to reduce large air currents from circulating round from the hot plate to the cold glass, and two, to reduce the re-radiation coming from the black plate. As the honeycomb is made of glass or plastic the incoming rays are not restricted. One loss it can cause is by conducting heat from the hot plate to the cold glass. The reason for the gaps now becomes apparent in that they interrupt that loss path.

In the second arrangement, another form of honeycomb is formed into what is known as a 'vertical plate' collector. While this is often formed by flat black plates that are at right angles to the main plate, another form is constructed from black round cans. Chapter 14 shows this in detail and explains the operation.

The stage has now been reached where we have a complete watertight collector *plate* that has either been purchased from a manufacturer or fashioned by your own hands. To complete our collector *unit,* we have to fabricate certain other simple parts and assemble them along the lines shown by Fig. 88

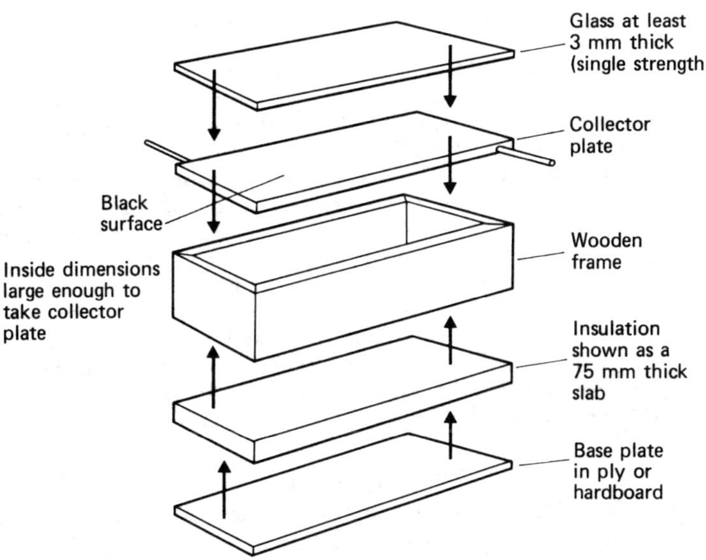

Fig. 88. The simple parts that make up into a complete collector unit.

Notes

1. All parts are asembled on the wooden frame made of 25 mm. (1 in.) by 125 mm. (5 in.) planks.
2. The collector plate is fixed inside the frame so that its black surface is 25 mm. below the top of the frame.
3. The frame is turned over to fit the 75 mm. (3 in.) layer of heat insulation. If fibreglass insulation with an aluminium facing is used, the facing goes towards the collector plate.
4. If a loose insulation such as sawdust (yes, a good cheap answer) is used, light partitions should be added to reduce its movement. If sawdust is used, all electrics have to be out of the box to negate any fire risk and even then it may well contravene Building Regulations.
5. The base plate can be very light-weight as it only has to support the insulation. But remember that rain may creep under it and encourage rotting, a resin bonded ply or marine ply is indicated.
6. The glass fits down onto the frame. It is best held by cutting a groove round the top of the frame about 15 mm. wide, and then lining it with a rubber strip for the glass to sit on. A clamp strip then holds the glass tight against the rubber. For long trouble-free service, putty and other caulking compounds are not recommended; they also preclude the possibility of easily removing the glass.
7. If it is preferred to have the glass in two pieces, take care that the joint is waterproof. Either lap it so that the water runs *over* the joint and not into it, or, use a glazing bar fitted with gasket strips along each edge of the glass.
8. If you are opting for a double glazing, these are the differences. The basic frame has to be increased in depth from 125 mm. to 150 mm. and the collector set down to 50 mm. and not 25 mm. Then by putting a 25 mm. by 5 mm. wooden strip round the inside of the frame on top of the collector plate, the extra glass can be rested on it. Or, if plastic sheet is used, it can be fitted to a 2 meter by 1 meter by 25 mm. frame and dropped in to sit directly on the collector plate (the plastic ending up 25 mm. from the plate as did the extra sheet of glass). In either case, the original top sheet of glass is fitted as previously described.

Up until the Los Angeles World Solar Energy Conference of July 1975 (where the author was a delegate), all but a very few imaginative people considered that a flat sheet of glass or plastic was the only cover for a collector.

Quite incredibly, three inventors, one from Europe and

two from the U.S., independently came up with the same alternative – a layer of clear fluorescent light tubes! Due entirely to the tubular section, great strength is given to the glass, even with a considerable reduction in its thickness. Consider the advantages:

1. Tubes can span 2.5 meters (8 ft.) without center support.
2. To cover the same area they only weigh one-fifth that of glass.
3. To cover the same area they only cost one-fifth that of glass.
4. A missile will only break one or two tubes, and they can easily be replaced without disturbing the complete assembly.
5. A single layer of tubes is nearly equivalent to double glazing.
6. The possibility of evacuating the air from the tubes exists, to further improve their performance. (This is a specialized undertaking involving some risk.)

One problem that is being worked on is the gaps that appear between the tubes, but even without this being solved the idea is still very good, having been used in the collectors we have discussed, and the 'Thomason' type to follow.

Supplies of tubes can be obtained from lamp manufacturers as clear open-ended tube of any reasonable length. Alternatively, they can be re-cycled from old tubes, the phosphor is easily removed by pulling through a damp tissue, being careful not to inhale the dust. With re-cycled tubes the glass cost is down to nil! Tubes should be mounted on a 'spring-in' arrangement by butting each end into a pillow of foam or E.C.H. plastic.

Thus far all the collectors we have discussed in detail use *water* as the medium to transfer the collected heat to the point of store or use. The option we now describe uses air as that medium.

The collector is a flat plate device, with a single glass cover. It is air operated and has minimal air storage (i.e. suitable for temperate and hot climates), with a home constructed plate. The plate is not considered as a separate item this time, the collector is taken as one complete unit. See Fig. 89 for the general arrangement.

The now familiar glass topped box appears again; it lets heat in the front, and the usual insulation at the rear stops it from

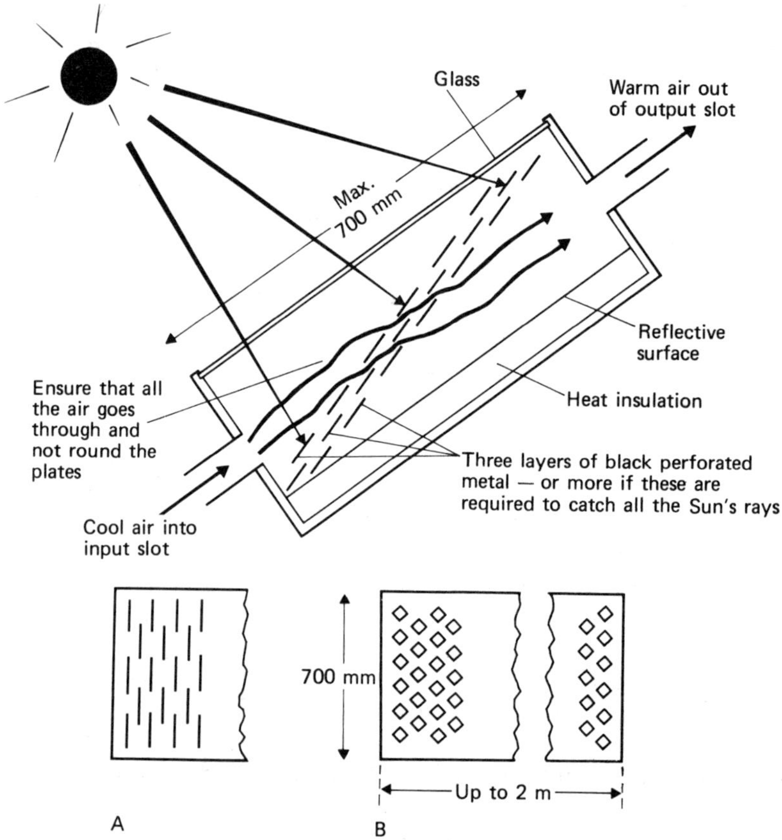

Fig. 89. An air operated solar collector. The slots extend the full length of the collector – up to 2 meters.

getting out the back. But the heart of this collector is three perforated metal plates spaced slightly apart and arranged in a particular way. Practically any type of perforated metal will do – copper and aluminum with their superior conduction properties are best – as long as the sheets are angled across the box and meet these two requirements: (1) at whatever angle the Sun shines on the assembly of three or more plates, it does not shine through the plates and onto the insulation at the back. (Not that this will hurt the insulation, but rather if it goes between the plates, some of its energy is not caught. And (2), the air flow through the box passes via the perforations in the layers of plates by a somewhat 'torturous' path. In this way the air extracts the maximum heat from the plates.

There are several variations of the plate that come to mind quite readily as they only have to meet the previous criteria to work. One idea is to slit heavy grade kitchen foil as shown at 'A' in Fig. 89, and then pull it out, as in 'B', to make your own perforated metal. It is, of course, painted black.

We have now completed the section dealing with the production of a complete collector *unit*. Alternatively, you could have arrived at this point by purchasing a complete unit ready to fit to your system. But before we leave collectors, we have to describe another quite different collector concept. It is an arrangement that is eminently suited to temperate climates, although as with most other temperate climate designs, quite suited to hot climates also. We referred to it previously as a collector that, was 'an integrated part of a roof'. It is an excellent large collector that forms the basis of what are very likely the best and most well-known Solar houses in the world. These are the Thomason houses – invented, designed and built in the Washington, D.C., area by Harry Thomason, an ex-patent attorney, now fully engaged in Solar work (and regarded as the Maverick of the U.S. Solar scene). It is worth noting that Harry Thomason, while never part of the 'laboratory fraternity', has by his own self-funded efforts achieved these extremely solid results. There is now a U.S. Government funded project underway to evaluate his work, and if past experience is any guide, no doubt the evaluation will cost more than Thomason actually spent doing the work.

A story is told that Thomason's idea to make a collector by running water down a hot roof came to him one day when he was sheltering from a summer thunderstorm. Instead of cold water dripping down his neck, he noted that it was *warm* water. It is from the observation of such apparently insignificant differences that great ideas are born.

There are at least two installations in England which use a similar arrangement.

With this concept of an integrated roof collector, one has generally to accept that this is going to be a 'do-it-yourself' operation, or at least that you will have to have it done for you. Whilst it is now possible to purchase ready-made units, the original collector really is the roof. Clearly, it requires somewhat more courage to tackle one of these installations although one of the first components to go back onto the roof is metal decking – which makes it quite waterproof again.

This is the arrangement, and we are discussing the version where the collector is built onto the roof *instead* of the tiles. It takes little imagination to envisage the alternative arrangement (which obviously costs more) with the tiles under the collector. Leaving the tiles on gives fixing problems but does leave an umbrella if things go wrong with the collector. Refer to Fig. 90. The top sketch shows a roof being fitted with a collector instead of the usual tiles. The rafters are exposed and the first part of the collector is going on at the bottom left. While we have shown a conventional roof, a house specifically designed for this type of collector often has a roof extending nearly down to the ground on the Sun side. The remainder of Fig. 90 shows a cross-section of a completed part of the roof. The version shown (without the tiles) is the most complex method, and we discuss a simplified version later.

A collector of this type will normally cover one side of a roof with the buildup starting when the roof is at the stage of having bare rafters only on the side to be fitted. This can be when the house is being built, or by stripping off the existing tiles, etc., back to the rafters.

The actual buildup starts by covering the roof with a plastic sheet, lightly tacked onto the rafters, to act as a vapor barrier and stop damp from spoiling the insulation. Next comes a plywood sheet about 10 mm. thick to eventually hold the insulation in place. Then, onto the plywood go wooden spacers which

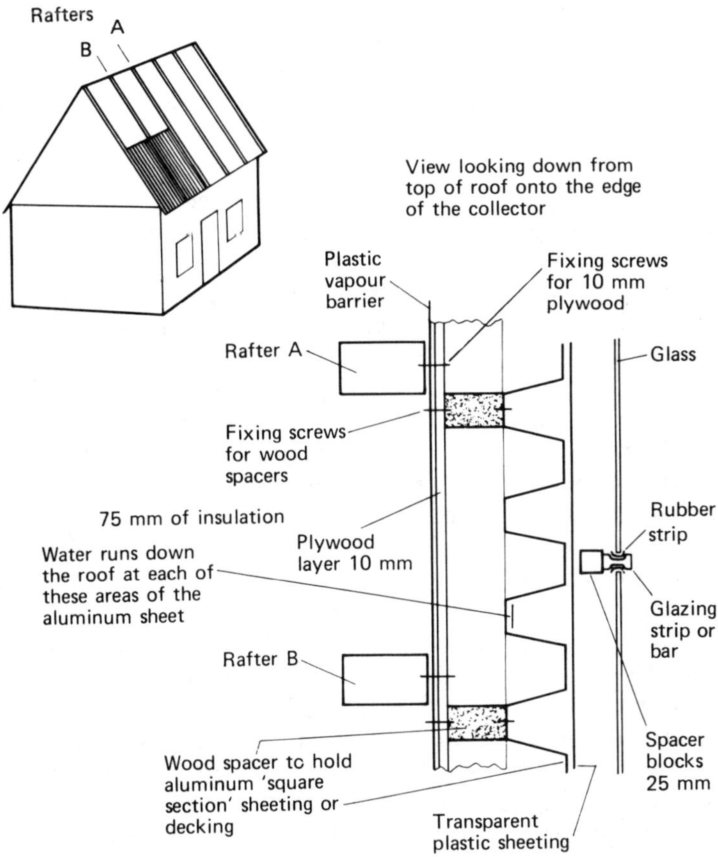

Fig. 90. A collector that covers the roof instead of the tiles.

will eventually support the aluminum sheet. They will not necessarily be next to each rafter as we have shown them, as they have to coincide with the corrugations of the aluminum. For reasons of strength, however, they should be close to the rafters and if set just to one side (as shown) the fixing screws can go through the 10 mm. ply *and then* into the wooden spacers. The space in between the wood spacers is then filled up with 75 mm. of heat insulation to virtually cover the complete roof area. Now comes the aluminum sheeting. These are readily available aluminum decking sheets about 2 mm. thick. Thickness is not very important (the thicker the better if cost was no problem) but it is well worth trying to obtain *embossed* sheets. This is a manufactured roughness of the sheet's surface which helps to spread the water over a larger area and makes it more efficient in picking up the heat. The aluminum sheets are screwed to the wood spacers, as shown, using a small amount of filler in each screw hole to keep water from entering and eventually rotting the wood. (An alternative arrangement worth considering is to make the wood spacers so that they come to the *top* of the aluminum sheet. This way the fixing holes are not in the waterways.) The application of the black surface then follows; assuming you have not been able to get your sheets anodized or painted black. Next, comes the glazing. The full system we are going through uses our rudimentary arrangement of double glazing by first putting a sheet of clear plastic over the aluminum sheet. Further spacing blocks of about 25 mm. thickness follow, and effectively hold the plastic sheet in position. Onto the top of the 25 mm. spacers go the last of this multi-layered sandwich – the glazing bars. The glass slides into these in the prescribed manner for the particular type of bar you have chosen, but whichever one it is, use a rubber (or similar) seal where the glass meets the bar.

We now go through a simplified version of the main part of the collector assembly before discussing the top and bottom of said assembly. These are the parts where the cool water is fed in and the hot water taken out.

This simplest version starts by screwing the aluminum (preferably anodized and kept free of scratches) sheet straight to the rafters – that's a good beginning! The usual insulation is then fitted between the rafters and held in position by 'best methods available'. Then if not already done, the aluminum is blackened. The glazing strips are fixed straight onto the top of the aluminum and the glass fitted to the strips. That's all!

Obviously, the intermediate degrees of complexity between the two extremes we have outlined exist, but, and this is very significant, even our first and most complex version of this collector is (at the present) the *cheapest collector* per square meter we know.

As you may have realized by now, the operation of these collectors is by simply trickling water down the corrugations of the aluminium sheet, to achieve this we must add the means to feed the water in at the top – and add a gutter to collect it hot at the bottom. The top feed is a simple 15 mm. copper pipe going across the apex of the roof with 1 mm. holes at each waterway to squirt water into each one. (It is prudent to try the water flow before putting the glass in position.) To collect the water at the bottom of the aluminum sheet requires the fitting of a simple gutter arrangement. Slope it in the way of ordinary gutters, thus quickly getting the hot water out at one end. Keep the gutter inside the glazed area in order to protect it from the cooling winds. Back at the top of the roof, the very highest glazing strip of the rows that go across the roof (glazing strips go up and down the roof as well as across it) is fitted with flashing that goes up the last 50 mm. or so of the roof, over the top and down the other side, to be nailed onto the roof on the other side. The following notes apply to all the preceding discussions.

1. So far 'glass' has been used as a somewhat vague term; let us expand it a little. The old method of grading the thickness of glass by weight – an extraordinary concept – has been replaced by the thickness method making it clear to all and easy to check (double strength being 1/8 in.).
2. Buy horticultural grade glass. Although it has a certain number of ripples in it making it unsuitable for windows, it is just lovely for collectors.
3. Avoid 'green edge' tinted glass if possible as the impurities in it restrict the Solar rays. White or blue edge is the best if it can be obtained.
4. Make your glass easy to get out. This, plus the expansion and contraction of the surrounding materials makes caulking compounds a poor long-term solution.

5. Plastic film in place of glass can be considered and is used in some commercially available collectors, but don't look on it as an everlasting material such as glass. The problem with plastic is that long exposure to the ultra-violet rays tends to produce brittleness (then tearing) and also cloudiness or loss of transmittance (of the rays). Various glazing products are available, and new and better ones are appearing, with a life up to 20 years.
6. The wooden frame round your collector should be well protected from the weather both inside and out, by several coats of good paint. One idea is to paint on resin, adding a thin sheet of fibre-glass and finishing with more resin – that really holds the weather.
7. A small series of 'weep holes' at the bottom of the collector to allow odd accumulations of water to run out are useful. They also allow the box to 'breathe'.
8. The angle at which to mount a collector (by angle we mean the number of degrees it is up from being flat on its back on the ground) is the subject of much discussion but boils down to this: If it is mounted at a similar angle in degrees as your location's latitude (also in degrees) it should give the best *year round* results. (This is based on the fact that at midday on the *middle point* between the Sun's highest and lowest traverse, the collector will face directly at the Sun. This actually occurs twice a year on the so-called equinoxes of 21 March and 23 September. But, in practice, it is better to add at least 10° and more *if you wish to trade a small winter* increase for a large summer loss – see the map (Fig. 52) again. But keep the whole thing in perspective – in the final calculation the existing roof angle appears to surpass the power of the computer!
9. The orientation of the collector is slightly better at 15° West of South in the Northern Hemisphere (15° W of N in Southern) because the Solar day is generally slightly delayed as the morning haze is burnt off. Fifteen degrees West of South is found by positioning the collector to face the Sun at 1 p.m., but if your area has 1 hour of daylight saving your clocks will show 2 p.m.
10. Try out your collector for as long as possible on the ground – or at least in some easily accessible spot.
11. When you mount it, mount it very firmly. Think of those gusts of wind that might get under it. One bolt right through a rafter with a nut on the other side is worth a dozen screws.
12. Collectors can be fitted crossways on a roof, if so it makes it easier to increase the angle of them, relative to the roof, by lifting up the top edge.
13. Collectors can go onto porches, lean-to-roofs, gable ends, or as canopies to windows, or verandas. Or just leave it on its own

against the house.

14. A 2-meter by 1-meter collector might well weigh about 40 kgm. (88 pounds), which is about as much as can be double-handed. If you are single-handed, a 1-meter by 1-meter size is indicated.
15. Remember that polystyrene foam insulation will melt if left in full sunlight inside a waterless collector.
16. Insulation properties vary so much that they should be carefully checked, i.e. Urethene foam has only half the loss of glass wool.
17. Think where the Sun gets to around your house in the winter, if you are fixing a collector in the summer. The difference from maximum elevation on 21 June to minimum on the 21 December is 47°; add half of the 47° to 90° less your latitude to give the highest number of degrees that the Sun will ever get to over your abode. i.e. at 40° lat. 90°–40° = 50 3 23½ = 73½° maximum.
18. But more down to earth; check whether you require planning permission or violate any local codes or building regulations or invalidate any insurance cover before you erect any outside structure.
19. A good harvest of cheap copper pipe and thermal insulation can often be found on junk refrigerators, especially the larger variety.
20. In systems with mixed metals use a 'galvanic isolator' at the point of contact to avoid corrosion. This means the insertion of a piece of (say) plastic tube so that the two different metals are not in direct contact.
21. Flow rates of water should be kept below 1.2 meters per sec. (4 ft/sec) with aluminium or copper systems to avoid excessive internal wear due to rushing water.
22. Ensure all flux is removed from copper systems and that the external surfaces of aluminum components are painted.

To conclude, the present state of collector designs leaves this intriguing thought to ponder:

Collectors providing hot water for home use are about 60% efficient. In the unlikely event of a researcher skirting round several laws of physics and coming up with a near perfect collector, he will only have made it 50% more efficient – the 600 watts output rising to 900 watts – and if this is achieved with *no increase* in price, there will have been a very creditable gain. But, as there is little or no limit to the *area* from which Solar Energy can be collected, the man who designs a collector (still only 60% efficient) but at *half the price* makes the more valuable contribution. By covering twice the area for the same total cost, the output now rises from 600 watts to 1200 watts.

CHAPTER 10

Solar Heated Domestic Water — 5 Schemes

Very few people can reasonably expect Solar to *completely* replace their existing heat sources. In temperate climates it will get very close with domestic water but less close with home heating. Typically, one is doing well if 75% of the hot water and 50% of the heating is obtained. In hot climates it should provide 100% of the hot water and 75% of the heating.

As Solar has to be looked at as a partial heat supplier, it presupposes that there will be some other form of heating device available. Our next stage is to explain the various possibilities for feeding in the Solar heat, thereby reducing the demand from the conventional source.

The 'adding on' of a Solar unit does raise certain problems and alternatives along the way that you may, or may not, find acceptable. By talking our way through the various stages, the multiplicity of answers and options available can be seen in their correction context and thus better relate to *your situation.*

A typical installation that might exist in your home to heat domestic water is shown in Fig. 91. We see that a hot tank stores water that is either hot water coming in from a boiler via a heat exchanger coil BB, or from an immersion heater D or both. As water is consumed from the tank (not circulated in a captive circuit as with radiators) it follows that a replacement supply of cold has to come in, as at G. The system is completed by a Thermostat E which regulates the immersion heater and maintains the water at the chosen temperature. The same applies to the thermostat controlling the boiler.

From now on, we are talking as if there is only one thermostat, and that is the immersion heater one. And it is there, at the thermostat that the first problem occurs, and where the first alteration to your existing system occurs. The alteration? To merely turn the thermostat down to a level where the hot water is at the lowest acceptable level. A level where little or no cold is added to make it useable – say to 50°C (122°F). This simple alteration has already started to save your fuel as the losses from your hot tank are less if it is kept at a lower temperature. (Recall how in Chapter 7 the heat loss through the house walls went up with increased temperature difference between inside and outside? It is exactly the same for the temperature difference inside to outside the wall of lagging round the tank.) The appreciation and actioning of this one point will more than repay your investment in this book! Now for our first scheme.

To merely erect a Solar collector and feed its outflowing hot water straight into the hot tank at G in Fig. 91, would not be very effective. Assume we put the Solar heat in and switched the immersion heater *off* – we would get hot enough water in the summer, perhaps up to 60°C in temperate, and 70°C in hot climates; but in the winter it would be more like 30°C and 40°C. Not quite good enough. (For those who see a way that they might use the arrangement, this is the additional data they will now require – it also holds good as general data for all to use.)

1. The collector should be 1 square meter in size for every 50 liters of water in the tank. (Work out your tank size thus: knowing that a 1000 cc. or cubic centimeter car is a 1 liter, then a tank measuring 100 by 40 by 40 centimeters is 160,000 cc. or 160 liters.)
2. The stated collector size will produce one tankful per day at 60/65°C in summer and 30/35°C in winter in temperate climates; or 70/75° and 40/45° in hot climates.
3. An average family of 4 uses 40 liters of hot water a day per person, rising to 50 liters each for a 2-person household.

To continue. By connecting a collector to feed into the hot tank at G and returning it to the collector via F – and observing

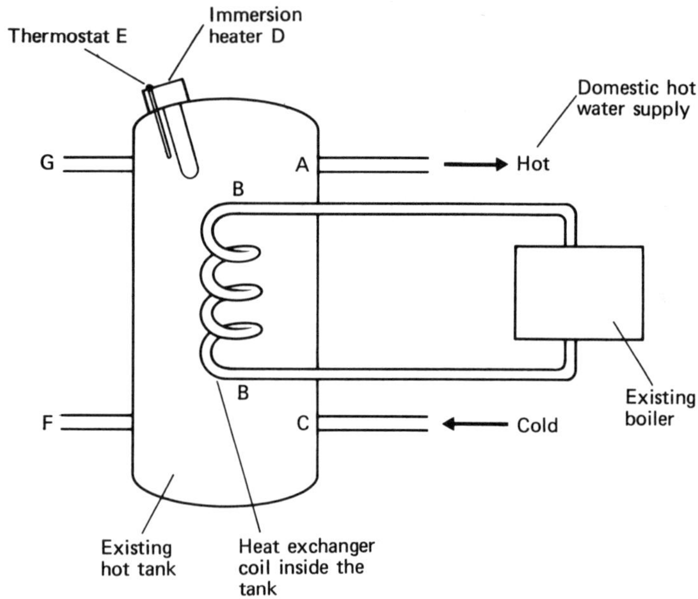

Fig. 91. A typical domestic hot water system. The only alterations are the entry and exit points G and F that allow our 'hot water scheme 1' to be added.

the general requirements of the thermosyphon system – we could get some results with the immersion heater off. But supposing it was *on?* Now the heater would switch on whenever the water dropped just below its 50°C setting and the Solar heat excluded until it *exceeded* 50°C – not good as that really restricts it to mainly a summer contribution.

Therefore, by attempting to use the existing hot tank you get hot summer water and warm winter water when the immersion heater is *off:* or, summer Solar heated water and no winter Solar contribution with it *on.* Actually, there is one little trick that makes better use of the immersion heater. By wiring it via a timing device, it can be arranged so that the heater can only come on at night – thus giving the best results of the three ways discussed. If a time switch is used, the immersion heater should now be fitted half-way down the tank, whereas if it is available for use at any time, it should be limited to warming the top third of the tank only.

To surpass all three, our second scheme in Fig. 92, shows that a better answer is to use a second tank – and in many houses, it is already up there in the loft as the coldfeed tank.

The idea of using the second tank is use it to store Solar heat, leaving the collector to get it as hot as it can, any time it can.

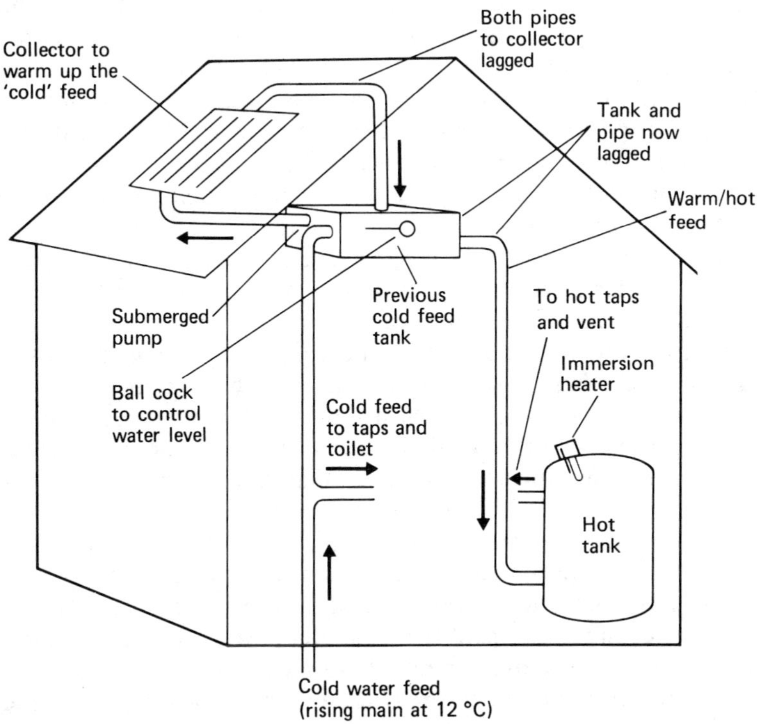

Fig. 92. The use of an existing 'cold feed tank' to store solar heat. Hot water scheme 2.

The resultant hot water (35° to 65°C) is then used to feed into the hot tank *instead of feeding in cold* (12°C). The heater is now only required if the incoming Solar water is below the 50°C setting of the immersion heater.

(It is interesting to consider how cold is cold water? Think back to Chaper 5 and the storing of heat in the ground. Recall that at a depth of about 2 meters the temperature is always about 12°C – 2 meters is also about the depth that water pipes run at – so that's just about the temperature that the water comes out, and why it still feels cool on a hot day.) Now we can see that even if our store of Solar heated water is only going into the hot tank at 35°C, it has still added 23° to the original 12°C – more than half-way to the required 50°C. This is real progress, and yet we still have other ways of using the second tank. But before even getting involved with our first scheme using a second tank (hot water scheme 2), we must go into the generalities of moving hot water from the collector to the tank.

Thermosyphon or pump? The former is a delightfully simple method of achieving the desired result. It uses no extra components that might go wrong, is cheap, completely automatic and is working well in thousands of units in Australia, Israel and other parts of the world. But these are its three limitations: the tank has to be *above* the collector; it is not easy to protect against frost damage; it has limitations when trying to add it to an existing system. Obviously, it is a sound system – try it by all means – our lack of further system explanation reflects its inherent simplicity and the availability of many excellent commercial units rather than disapproval.

All systems discussed, including the first, are pumped systems – the pump being a little device similar to those that pump water around central heating systems. But as pumps are related to the frost problem, we deal with that also before returning to hot water scheme 2. In a pumped system, by carefully choosing the collector type, and judicious runs of pipework, one can ensure that the water *drains out* of the collector once the pump has been switched off. Emphasis is given to draining the collector, as being outside, it is both vulnerable and costly to damage. In your own system make sure that the collector itself is self-draining; note that the one shown in Fig. 87a, is *not* self-draining, but by rotating it 90 degrees and slightly angling the pipe runs, it can be made so. Pipes to and from the collectors and tanks should be run with a deliberate gradient to them, or 'rise' as they call it in the trade, in order that they too will drain. Actually, the pipes will have further protection as they are wrapped in insulation and located mainly inside the roof.

Two further methods of protecting against frost damage are known if the collector cannot be drained. One, is to periodically pump around warm water when frost is detected. Two, is to include a small immersion heater in the collector. While both methods are at first slightly wasteful, they should be weighed against the losses associated with a heat exchanger, it losses are accountable all the year!

With a pump to move the water and our frost problem solved, we can now return to scheme 2, our first of four methods using the 'cold feed tank'.

The tank's original purpose was to store cold for *various home uses,* with cold water coming up from the ground at 12°C to fill it. What we are aiming to do is to turn the cold feed tank into a 'warm/hot feed tank' by putting in our Solar warmth.

Now, instead of re-filling our *hot tank* with cold water we refill it with warm or hot water – just what we aimed to do, and both of the limitations of hot water scheme 1 are overcome. A snag would appear if the original cold feed also fed into cold feeds or the toilet; they would now receive Solar heated water! In Fig. 92, as we have shown all other cold feeds coming from the rising main there is no problem. It is possible that your installation is that way now, or can be altered to be as shown, but if neither are possible Fig. 94 provides the answer. But before we go that far, there is a variation of Fig. 92 that is worth considering. This is our hot-water scheme 3, see Fig. 93.

Fig. 93 shows the same cold feed tank and the same connections to the hot tank, etc., but the piping from the collector is different. This time the water from the collector is taken

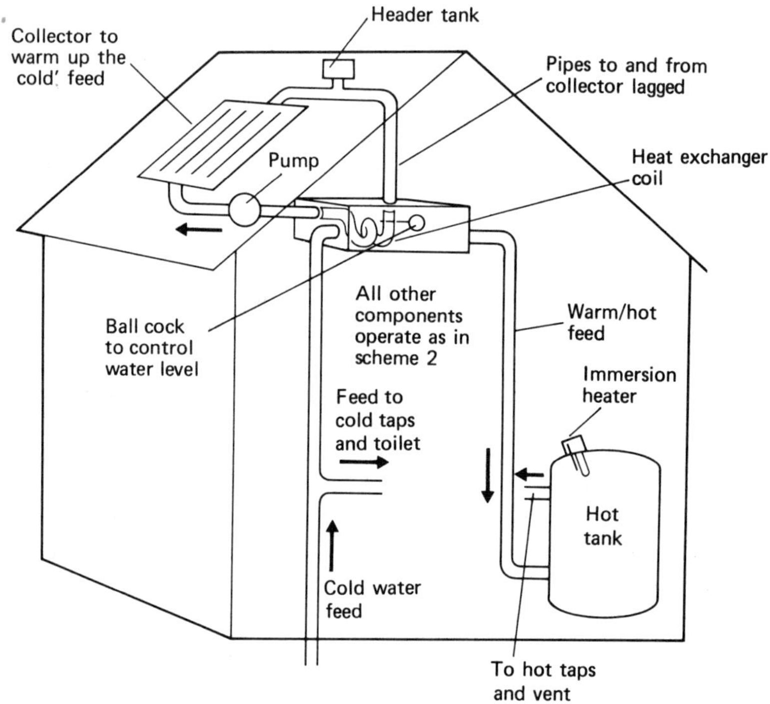

Fig. 93. By adding a heat exchanger coil the *same* loop of water goes round and round the collector circuit. Our hot water scheme 3.

through the cold feed tank via a heat exchanger coil. Fig. 27 shows a similar arrangement but that uses a finned heat exchanger whereas the more popular variety shown in Fig. 93 is made by coiling 15 mm. or 20 mm. copper piping. These devices can be purchased as separate units or as part of a complete tank assembly. The object of this apparent complication is to cure three problems. The first allows non-draining collectors to be used without danger from frost damage. The problem being solved by using an anti-freeze solution in a newly created 'separate water circuit' formed by the collector, heat

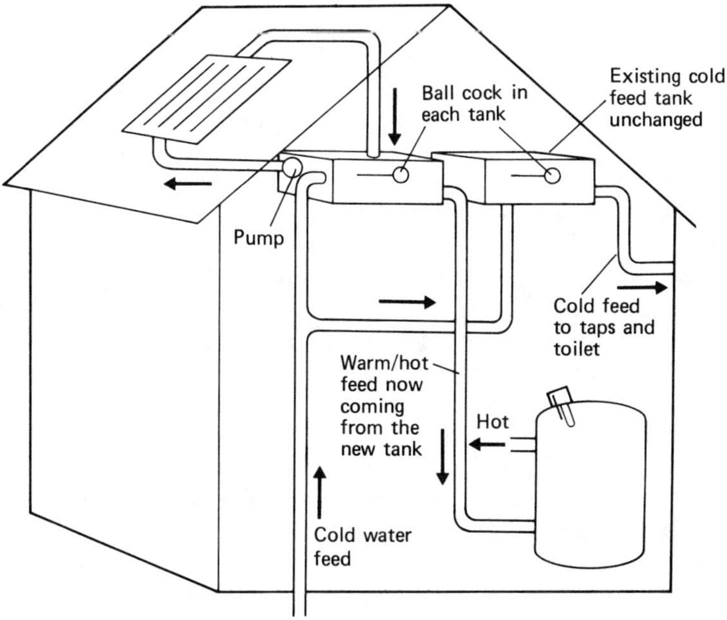

Fig. 94. This one is evolved from our hot water scheme 2. The added tank is on the *left*. Hot water scheme 4.

exchanger and small header tank (this can be a 25-liter plastic affair). The second problem can be furring-up due to deposit of lime, etc., from the water. Now, as the same water goes round and round the circuit, only a small initial deposit can occur. And third, is the reduction of corrosion in mixed metal systems. Here the theory is that in systems where the water comes in contact with (say) an aluminum collector and a copper tank, the less dissolved oxygen there is in the water the less corrosion takes place. The only practical way to achieve this is to make sure that the same loop of water goes around. It also helps to keep the oxygen content down if the *surface area of the header tank (open to the atmosphere)* is kept to a minimum.

An alternative to the header tank of Fig. 93 is to fit at the end of the short length of pipe, an 'accumulator'. This is a metal cannister, with a single water pipe connection, that will absorb (or accumulate) water inside itself as pressure is increased, and return the water as pressure is reduced. This way the expansion of the water in the otherwise closed loop is accommodated, furthermore, no water surface is exposed to the air to either allow evaporation loss or oxygen absorption.

Now to Fig. 94, our hot water scheme 4. This arrangement is a direct system, more allied to scheme 2 (rather than 3), the difference coming in the domestic cold feeds. It is necessary in those houses where the plumbing cannot be altered, to feed the cold taps direct from the rising main. In this case, they are just left as they are coming from the cold feed tank (and it stays as a cold feed tank) and another tank is added for it to become our warm/hot feed to the hot tank. The remainder follows from the diagram and the previous schemes.

Hot water scheme 5 is the obvious deviation from scheme 4 in the same way that scheme 3 developed from 2 – a heat exchanger is added to produce a 'separate water circuit'. Fig. 95 shows scheme 5 in full to complete our quartet of schemes.

With any scheme using a pump, some form of control is essential in temperate climates to switch it on and off. Constant circulation would eventually feed hot water to a collector that was not only not producing any heat input, but would then waste heat. This might happen when the Sun was behind clouds and would certainly happen at night.

Two possibilities exist for solving the problem. One uses a single thermostat at the *output* of the collector (normally the top); the other uses a pair of *matched* temperature sensing devices, one at the collector and the other at the tank that stores the water coming from the collector (the warm/hot tank in our previous 4 schemes).

The first arrangement is a straightforward, thermostatically controlled switch that goes ON at a pre-set temperature to activate the pump. The more subtle second arrangement measures the temperature of the collector *and* the stored water.

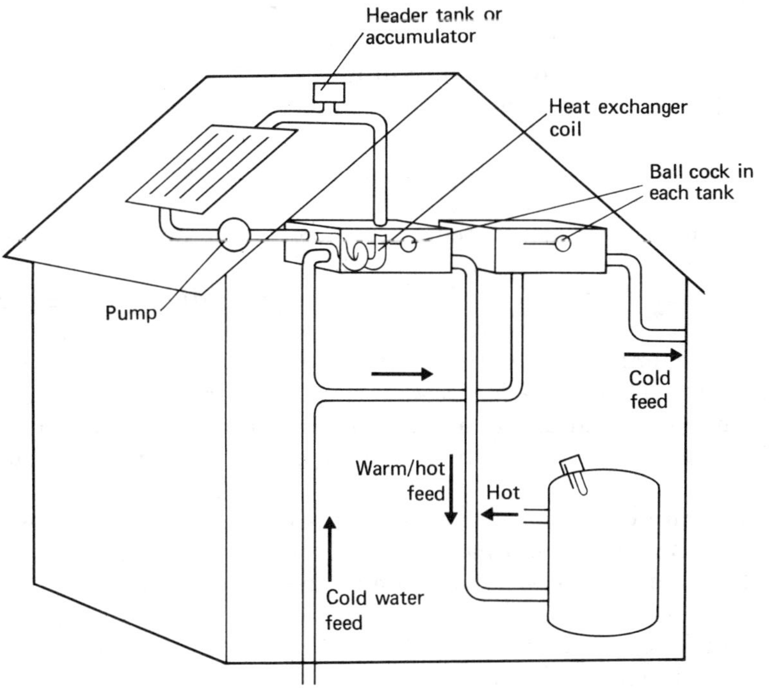

Fig. 95. By adding the heat exchanger from scheme 3 to the previous scheme 4, we arrive at our final hot water scheme 5.

By then comparing the two temperatures in a little box of electronics it puts out an appropriate signal to the pump to stop or go. If the collector is, say, 3°C hotter than the tank, the pump will be switched on. The beauty of the arrangement is that it holds over a wide temperature range, this allows heat to be added to the tank whenever it is just possible.

The electronic control boxes also come in an 'anti-freeze' version, they detect when the collector is nearly down to freezing point, at which time they close another set of contacts that can be used in one of three ways to protect the collector.

(a) Switch on a small water heater in the collector.
(b) Switch on the circulation pump, so using the tank's warm water to keep the collector unfrozen.
(c) Open an electrically operated valve to drain the water from the collector.

The Directory leads to suppliers of electronic control boxes and temperature sensing devices.

We have now considered five schemes that derive domestic hot water from Solar Energy. These five, plus the thermo-syphon arrangement should have given you sufficient coverage of the field to enable you to attempt your own installation – or to select one and have it fitted for you. The Directory again leads to names and addresses of suppliers, contractors and consultants who will accommodate all your needs from the supply of a pump to the designing and fitting of a complete installation.

For those who are carrying out all or some of their own work, the key to lasting success – and there is little success unless it does last – is attention to the mundane *detail of the work.* To conclude this chapter, we add a random list of small detailed points, any of which could mean the difference between success and failure.

1. Plastic pipes can be used to and from the collector – but if doubtful about their ability to withstand hot water, try a piece in boiling water.
2. A clear plastic pipe, even a small section, provides a good window to observe water-flow, and double as a 'galvanic isolator'. (This is not a complete isolator as cupreous oxide can still be borne through the pipe causing electrolytic reaction in say galvanized water tanks.)
3. Physically feel around your working system, as any hot points mean wasted heat. If you have only single glazing on the collector, this might be one hot spot and may convince you to double-glaze.
4. Keep pipe runs as smooth as possible to avoid air locks.
5. Relate your tank of hot water to your family size and your collector size to your tank.
6. Pumps should be chosen on the generous side and made of gun-metal or bronze. They can be either the 'let into the piping' type or submerged in the tank (low-voltage type).

7. On the last 4 schemes, ensure that the hot feed from the Collector *splashes down* into the tank, thus allowing air into the system when draining the collector to prevent freezing.
8. Any *closed* loop of water that is to be heated *must have* a point that allows for expansion. This means either a header tank, or an accumulator.
9. To guard against overheating your collector, you may wish to add a pressure relief valve in the top water outlet pipe.
10. Some collectors have an overheat vent at the top to relieve hot air and prevent boiling.
11. To aid the removal of internal scale, dirt ingress, fur, etc., one or two sludge cocks can be useful additions at the low points of the system.
12. If heat insulating lagging is not undercover, it should be wrapped in a plastic sheet.
13. Small air bleed points (as on central heating radiators) are useful in the system for the release of air locks.
14. Electrically operated valves are useful to drain awkwardly laid out systems. Switch them at the same time that the pump is shut down for the night.
15. After you have drawn your plans, get a plumber to look them over.

CHAPTER 11

Solar Heating — 4 Simple Schemes

Solar should be considered for home heating *after* you have converted your supply of domestic hot water to Solar. Domestic water demands are constant the year over, consequently, a Solar water heater earns its keep even during the summer. In addition, domestic water is generally produced (at least in the summer) by using the most expensive form of heat, electricity. It is clearly prudent then to invest in equipment that will not only be used the year round, but will also dispense with the highest heating cost first.

In proceeding, the assumption is made that some means exist to produce hot water to heat the home. If so, the Solar heating objectives now become: the use of Solar heat when *plenty* of Solar is available; the use of partial Solar, partial conventional if *some* Solar is available; the use of all conventional methods when no Solar is available – three distinct but related modes of operation.

Fig. 96 shows our first home heating scheme, a very simple arrangement, and yet one that is entirely automatic in operation. The unit itself is little more than our air heater (previously described in the chapter on collectors) with an extra airway formed underneath it to bring the input end back so that it is adjacent to the output. The unit is mounted outside the house, just below a slot that has been cut in the wall of the house. It may be possible to avoid this cutting, by either fitting the unit just below a window, or, by modifying a ventilation point (air brick) that exists in a wall.

The variations and constructional detail of the unit are left to suit your own requirements, but these notes will help:

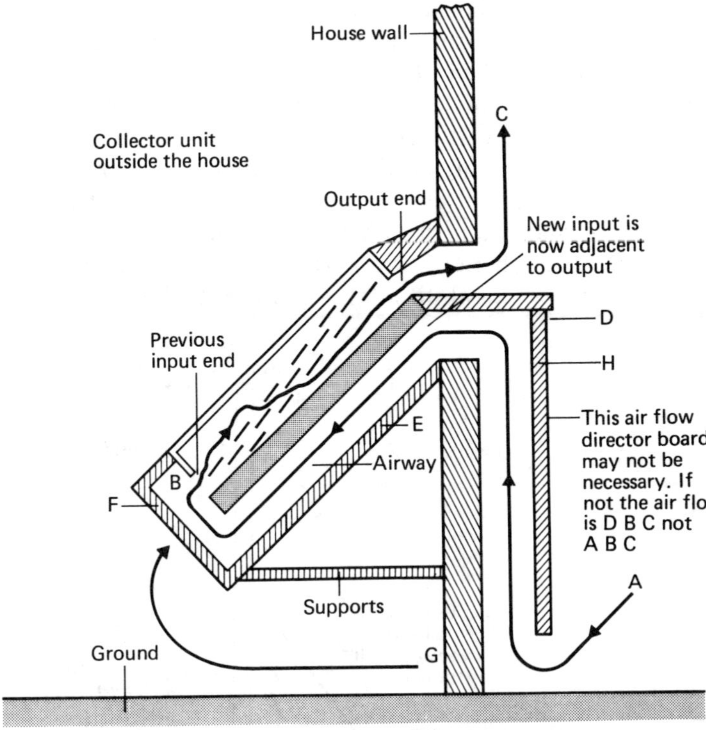

Fig. 96. Home heating scheme 1. The simplest solar home heating unit. Cool air at floor level A is drawn into the air heater at B to collect its Solar warmth and come at C. It is the rising warm air at C that causes the vacuum in the heater which draws the air in.

1. As the cool air from the room is effectively going outside the house to be heated, heat insulation is required under the airway at E & F.
2. The thermal efficiency of the unit is improved, if protected from cross winds (actually this helps any collector). One idea is to make the supports triangular in shape so that they block off any through drafts going under the unit.
3. Judiciously placed reflective sheets (silvery plastic?) can increase the effective size of the collector by up to twice.

These can be on the wall above the unit, on the ground beneath it, or on both, and they are generally angled to be at their most effective in the winter months.

4. A modified version of scheme 1 can be produced by cutting a second slot in the wall at G. This slot allows a shorter route to be made to the input of the heater at B, and a smaller hole formed at the original cut in the wall. The boards E, F, & H are then eliminated and the air flow is G, B, & C.

Our home heat scheme 2 is a natural derivation from scheme 1 and has so many possibilities that we are deliberately not re-

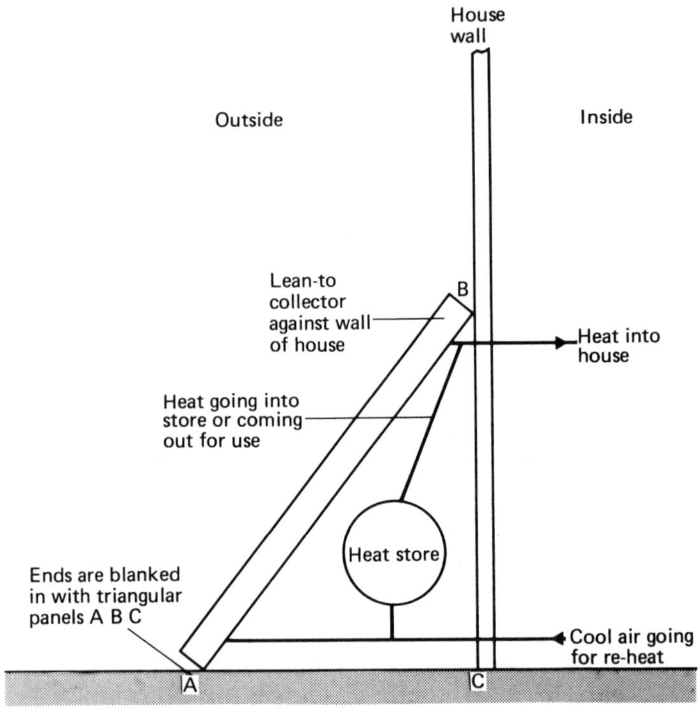

Fig. 97. Home heating scheme 2. More of a concept than a specific scheme. The lean-to collector is easy to install; unobtrusive in appearance; conveniently houses the store underneath itself; and is very well placed at ground level to work on for easy access.

stricting your thoughts by being too specific. Consider it as shown in Fig. 97, and then think about the variations:

(a) The collector could be air or water operated (2) options.
(b) The store could be water, pebbles, chemical, or none at all (4).
(c) There could be wall or ground reflectors, or both, or none (4).

These three groups of options alone give 32 possible systems *without* considering all the different variations of water collectors. There is truly great scope for the innovator here and all are schemes very conveniently placed at ground level!

The third of our simple schemes is shown in Fig. 98, and with it we return to more familiar ground as water and not air is used in the collector. The system comprises a collector,

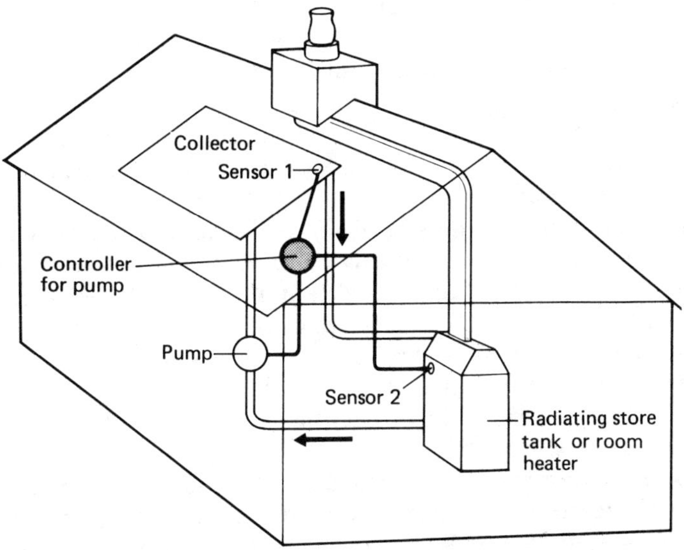

Fig. 98. Home heating scheme 3. The simplest way to get solar from your roof to warm your home. The disguised radiator is in the old chimney block.

a radiating store tank, and a circulating system; its objective is to independently supply heat for the home. It has no link to the domestic hot water system, and only an indirect link to the conventional heating system. (When the Solar system has warmed a room, the thermostat controlling the heat for that room senses that it is warm enough and turns the conventional heat down. This is an indirect link between the two systems.)

Scheme 3 operates as a basic pumped collector system with two sensors controlling the pump (as explained in Chapter 10) with all the heat going into a tank of water that is itself the store *and the room heater.* That's all! Getting the best from the system requires that the installer, not only disguises the radiator store, but also sees it is centrally sited on the ground floor (or lower). If this is achieved, the scheme takes the maximum possible heat from the roof, stores it, and uses it right in the center of the living area. Fig. 98 shows the tank put in the place of an old chimney. Other ideas include a flat tank in a recess, a long tank under the floorboards, a tank in a cupboard, a tank in a basement, a triangular tank in a corner (or two), a long tank doubling as a window seat, or, any combination of these. It is normally boxed in and insulated to retain the heat, and either a lever operated flap or a fan releases heat into the room as it is required.

We now consider the size of the collector and the tank store.

As one has to effectively live with the tank, let us assume an acceptable size and then ascertain if it is sufficient for our needs. One cubic meter is less than the volume taken up by an easy chair, and thus seems a reasonable size to consider. Remember from the outset though, that the water it will hold, 1000 liters, weight 1000 kilos (one ton) without the tank. This is the heat it will hold.

One cubic meter of water will take in and store 1 kw/hr. (our normal measure for a quantity of heat) by getting only 0.85°C hotter, a less than 1°C rise for every kw/hr stored. Conversely, for every kw/hr taken from the store, it would fall nearly 1°C. For those used to using 6 kw of heat in a room, the store supplying the equivalent heat would fall in temperature by 5.1°C each hour it was in use. (For those who are attempting to relate their requirements to an existing boiler rated in the old BtU. system, a 50,000 BtU. device is capable of giving out 15 kw. at maximum output.) The lowest useable temperature to which we can let the tank fall is 25°C, and this is our starting point for filling. By putting in 84 kw/hr we fill it to 96°C, which is just short of the maximum or boiling point. This is a very useful amount of storage. The 6 kw. rate could be used from it for 14 hours, or the equivalent of the full output of the 50,000 BtU. boiler for over 5 hours. But as it is only in summer that a tank temperature of 96°C could be expected, a more realistic top figure for the tank is 60°C. To get the water up to that from 25°C, it will have stored 41 kw/hr and will give back nearly 7 hours at the 6 kw. rate, or close to 3 hours at the full boiler rate. This is still quite a useful amount, all from 1 cubic meter; it may encourage one to look for another cubic meter! If one used the same volume of pebbles, instead of water, they will store (assuming 30% void space) only 40% of what water can. That is, the 1 kw/hr applied to one cubic meter would raise it 2.1°C, also, the weight of the cubic meter of pebbles would be double at 2000 kilos!

We now consider the size of collector required to fill the tank.

Chapter 8 shows that for each square meter of collector, you can expect so many kw/hr to arrive per day depending on your particular location. If collecting at the modest temperature of 40°C, the efficiency of the collector will be about 60%, so that if the graph shows that 4 kw/hr arrives each day, you will actually put only 2.4 kw/hr into the tank. If running at 60°C, giving more like a 50% efficiency, then only 2 kw/hr enters for each square meter of collector surface. Assuming you require 9 kw. to heat a room in the winter, and your location gives 3 kw/hr in these months, you will require a 6-square meter collector to merely put back the day's heat used. It checks this way: graph shows 3 kw/hr, but 50% efficiency reduces it to 1½ kw/hr/times, 6 for the size of the collector brings it up to 9 kw/hr per day. In order to gain on the day's use, and get some heat stored up for the sunless days, this bare minimum collector size (for the rather heavy use we chose) has to be increased. It is also advisable to allow *at least* 10% more in collector size for little losses that will creep in here and there, alternatively, leave enough room to add more collector sections at a later date. The general guide to ascertain maximum collector size is: the collector to be equal to one-half of the total floor area of the abode for houses in temperate climates; and one-third the area for houses in hot climates.

Another general guide relates to the size of the store tank. This time it is generally reckoned that 100 liters of water is required for every square meter of collector, in temperate climates, and 50 liters for hot climates. As a cross check and an aid to feel the situation, see how this now appears.

Consider a mythical house in Boston, Massachusetts of 75 meters floor area, and fitted with a 75,000 BtU. boiler. The graph for Boston shows that in January one can expect 2.3 kw/hr per day for a well-angled collector, so that with a 20-meter collector (under half floor area) it would contribute 23 kw/hr or 78,000 BtU. Quite sufficient. To give reasonable carry-over for blank Solar days, the store size would come out at 100 times the 20 or 2000 liters; equal to 2 cubic meters of water.

The mythical system described is really a top size system; which is just as well, as the tank is quite large and the collector size also gives problems. It is not easy to accommodate a collector that is nearly half the total floor area of a normal two-level house – think about it!

Simple scheme 4 is one worked out by a large Japanese property Corporation as the best method to provide individual Solar heating for apartments in Tokyo. Fig. 99 shows this to be a non-pumped or thermosyphon system. That is correct: the self-flowing system was previously used to provide hot water, and is now used to provide heat.

The basic idea is similar to scheme 2 in this chapter, but the particular geometry of apartments even allows us to dispense with using the pump. For those apartments without a balcony, the collector is fixed to the wall just below the window.

The same general idea could also be applied to a house if the collector was put a ground window level. Alternatively, the collector would be put outside an upper window and the collector/store on the bedroom ceiling or anywhere else in the room – provided it was above the top of the collector.

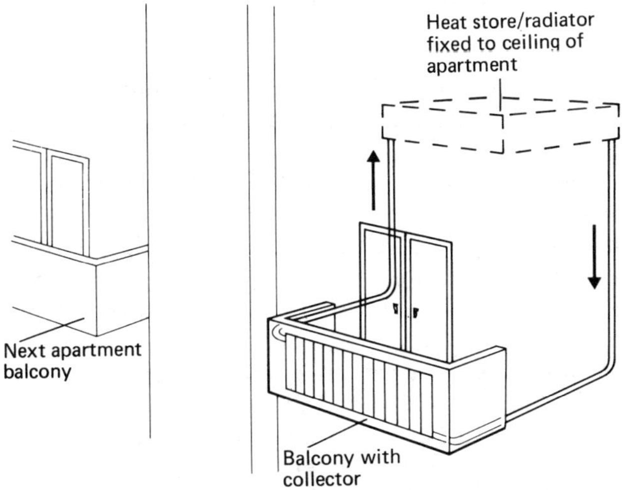

Fig. 99. The view into this Japanese apartment shows that as the heat store radiator is at ceiling height the heat from the collector on the blacony does not require pumping up.

It is quite apparent that there are many variations of size, and methods of execution to suit your particular requirements. Heating schemes 3 and 4 while being plain and simple, and causing no upset to any other house system, do have one other not so obvious attribute. Once the heat is in the tank the system is 100% efficient – the heat can only 'escape' into the house!

CHAPTER 12

Heat and Hot Water Schemes

Our scheme 5 provides Solar home heating, plus Solar domestic hot water and is designed to 'add on' to a conventional system that is already providing the same services. But to make the best use of Solar, the system will have to be operated in a slightly different way.

In this combination scheme, preference is given to using Solar to produce the home heat – as this is the largest user of the energy. But in tailoring the system to that use, it gives a different method of operation for this reason. Whereas domestic hot water is used once and thrown away, hot water for home heating is used over and over again. Water goes out to the radiators hot and comes back warm ready to be re-heated. It is this return of warm water that reduces the number of occasions that heated water from the collector can be used. The lower the temperature of the water coming back from the radiators, the more often the water from the collector can be utilized.

A second reason for having a lower operating temperature is that all collectors increase their efficiency at lower temperatures. A collector heating a swimming pool at 22°C (72°F) is close to 100% efficient, but swiftly drops to 50% if working up at 60°C (140°F). Collectors working at these higher temperatures are the ones that should be considered for double glazing, as it is the glass that is the hot spot where the efficiency is reduced. Certainly, if you are going to attempt to run scheme 4 at your present heating water temperature of 60/70°C (and you will see later why you might wish to do this), then double-glazed collectors are a near must, and even more so in temperate climates. To achieve the higher efficiency, and to enable the re-use of the warm water that is returned, scheme 5 is normally operated at below the usual temperature of 60/70°C (140 to 158°F).

Having seen the desirability of operating at 40/50°C, instead of at 60/70°C, this is how you raise the temperature of your collector should you desire to do the opposite and run at the higher temperature. There are 3 options. One, reduce the thermal losses, and this includes the use of double glazing; two, add reflectors to increase the amount of Solar; three, reduce the water (or air) flow through the collector. This last point comes about because the slower the water flows through, the higher its temperature becomes. A collector gradually builds up the water temperature by adding a few degrees each time the water comes round from the store. The slower it is allowed round, the greater the number of degrees the collector adds as the water passes through: 3° to 10°C being a typical increase. The collectors previously described can be expected to gradually raise the water to 40°C in winter and 90°C in summer in temperate climates; 50°C and 100°C in hot climates.

Therefore, while we recommend that scheme 5 be operated at 40°C winter, it could be used at higher temperatures if you wish to experiment. Fig. 100 shows this add-on system. To the left of the dotted line is the new installation, and this is basically the same whether the system is operated at the lower 40°C temperature, or the more normal 60/70°C. As the existing house radiator system is only altered if it is run as a 40°C system (details later), it might be a good idea to see if you can operate at 60°C before you alter the house radiators.

We now take you through the scheme 5 as a 40°C system, and, as you see it is in essence, scheme 2 of Chapter 10 – but with the cold feed tank now replaced by a much larger tank, as a Solar store, and another warm/hot feed pipe going to the boiler. Thus, there are now two outlets coming from the tank and a much larger Solar collector feeding in. The flow to and from the collector is driven by a pump, and it is activated by

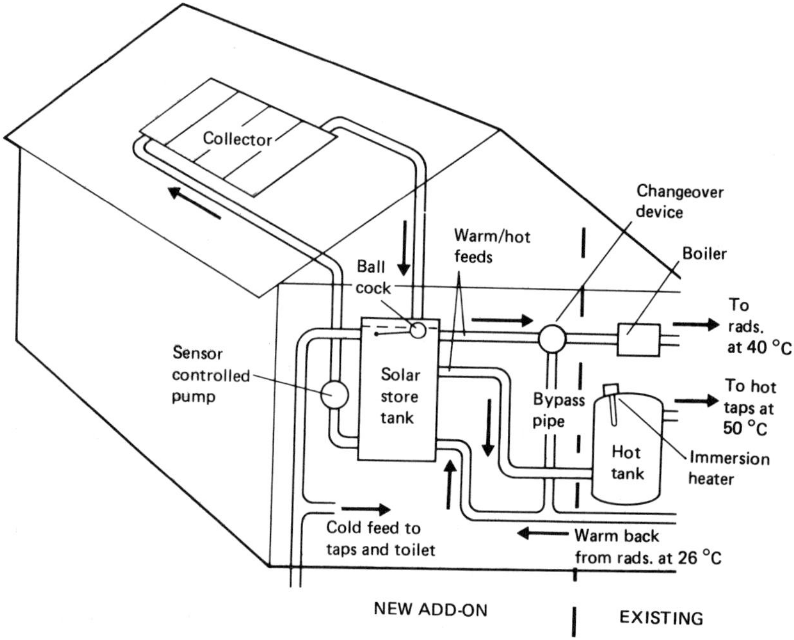

Fig. 100. Home heat and domestic hot water from solar. Heating scheme 5 adds on to an existing conventional system.

the usual sensors. Protection against frost damage is achieved by using the collector drain method, but care should be taken that all the precautions detailed in scheme 2 are carried out.

The two pipes, feeding out from the Solar store go: one to the hot tank, and two to the existing boiler. This later then circulates the water around the house radiators (not shown) and back to the heat store still warm at about 26°C.

The feed of domestic water has two modes of operation:

1. Into the hot tank at sufficient temperature for direct use – via hot taps.
2. Into the hot tank for boosting by the immersion heater before use – via the hot taps.

The feed of hot water for house heating has three modes of operation:

1. With the store tank at 40°C or morem the water goes straight to the radiators.
2. When the store tank is at 26° to 39°C, the boiler comes in to boost the water to 40°C before it goes out to the radiators.
3. When the store is below 26°C the 'changeover flow' device operates and bypasses the heat store. The operation of the boiler and the radiators is now a conventional system. Note: as the Solar store tank feeds the hot tank its high water point has to be above that of the hot tank by at least 500 mm. At the same time the store tank has to be low enough to allow the collector to self-drain.

To operate scheme 5 at 40°C requires that some compensation be made for operating the radiators at this lower temperature. There are three possibilities. Either the radiators have to stay on longer (generally not very practical or effective), or, more radiating surfaces are fitted, or, more output has to be obtained from the existing radiators by blowing them with a fan. While the first is barely worth considering, the second possibility does have a variation that is not quite as obvious as merely putting in more radiators (or the same size but double layered). This is to follow the same thinking as heating scheme 3. Use the radiating surface of the store as there can be no cheaper radiating surface than the one you have already paid for! The third possibility, the fan-assist idea comes in two forms. Either one mounts a small fan to get the warm air away from the existing radiator faster than it would move by normal convection, or the radiator is replaced by a proprietory device designed to do the same as the fan. Either method puts the output of the radiator up several times and should be adequate to compensate for reducing their operating temperature. Remember to distribute this larger cooler output as much as possible.

If the scheme just discussed does not suit your situation, three other solutions are possible by following the pattern of development used for the hot water schemes 2, 3, 4, and 5.

For example, if one cannot drain the collector, then the

'separate water circuit' of hot water scheme 3 can be applied to heat scheme 5 turning it into heat scheme 6. Alternatively, if the cold feed tank has to be retained, as shown in hot water scheme 4, this alternative can be applied to form heat scheme 7. Finally, by using the 'separate water circuit' and a separate cold tank, as in hot water scheme 5, we can produce heat scheme 8.

Whatever scheme you have under consideration you should follow these steps:

1. Examine the existing system piece by piece and pipe by pipe until you know it well enough to make a sketch of it.
2. Make a second sketch showing the modified scheme you wish to install or have installed.
3. Show them both to a competent plumber for approval.

There are many obscure reasons for carrying out certain practices in plumbing that are either 'tricks of the trade' learned over many years, or more likely, statutory regulations and safety requirements that have to be followed. On the other hand, plumbing is often presented as something of a 'black art'. It isn't really, as everything should be quite explainable to the layman.

CHAPTER 13

Thomason, Home Heating and Hot Water

The Schemes in this chapter are based on the Thomason 'Solaris' system; the colloquial 'water on a hot tin roof' system, very likely the most successful, viable, and time proven system in the world. With its quick reaction to short periods of Sunlight, it is one that is well suited to temperate climates, and yet in no way limited in its hot climate performance.

The basic elements of the system are shown in Fig. 101, there you see water being pumped *clockwise* around an integrated roof collector (details in Chapter 9), then into a tank store for home heating. In addition to the water store there is also a pebble store surrounding both tanks. The complete store can be visualized as a giant brick-built bin full of pebbles, with two tanks buried in the middle. The pebbles double as heat source and heat radiating surface to provide hot air to warm the house. Back-up for the system is provided by a boiler that also produces hot air.

This is the operation of the single arrowed water circuit of Fig. 101. Water is pumped to the top of the house and released down each of the multiple channels of the hot roof, thus collecting Solar warmth on its way down. The resulting warm water arriving at the bottom goes first into the pre-heat tank, where a heat exchanger inside it starts to heat domestic water, and then onto the store tank before completing the circuit back at the pump. This continual circulation of water (when the Sun is out) soon heats up the pre-heat tank, the top of the store tank, and eventually working its way down until all of the store

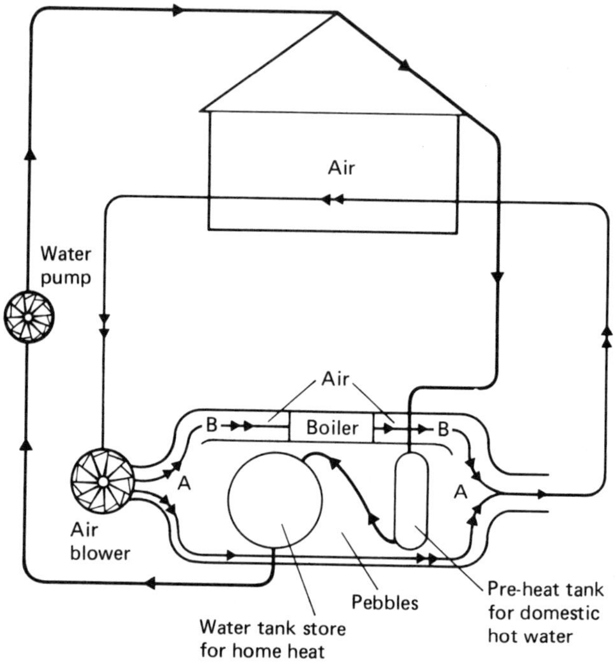

Fig. 101. The basic elements of a very fine system. The clockwise water circuit (single arrows) gathers heat as it comes down the roof to heat up the two tanks. The double arrowed anti-clockwise air flow gets its heat either from solar path A to A, or boiler path B to B.

tank is hot. Store tank temperatures of from 38° to 60°C can be expected in temperate climates and 45°C to 70°C in hot climates.

This is the operation of the double arrowed air circuit of Fig. 101. As the store tank is warmed, it gradually warms up the surrounding pebbles so that cool air blown across from A to A is sufficiently warmed to heat the house. This warm air is circulated through the house and eventually returns cool ready for re-heat and re-circulation. If the Solar heat from the pebble store is insufficient, the passage A to A is blocked off and the air passage B to B opened. The cool air passing through is then re-heated by a conventional air heating boiler and the air blown round the house as before. Thus we have two basic circulating currents: one the air (which has two options), the other the water. Now let us consider them in a practical system.

Fig. 102 shows heating scheme 9 as it has been used in a dwelling since 1959. A single-story structure with a 9-meter by 12-meter ground floor area and a basement of similar size. Nearly one-third of the basement is taken up with a giant brick

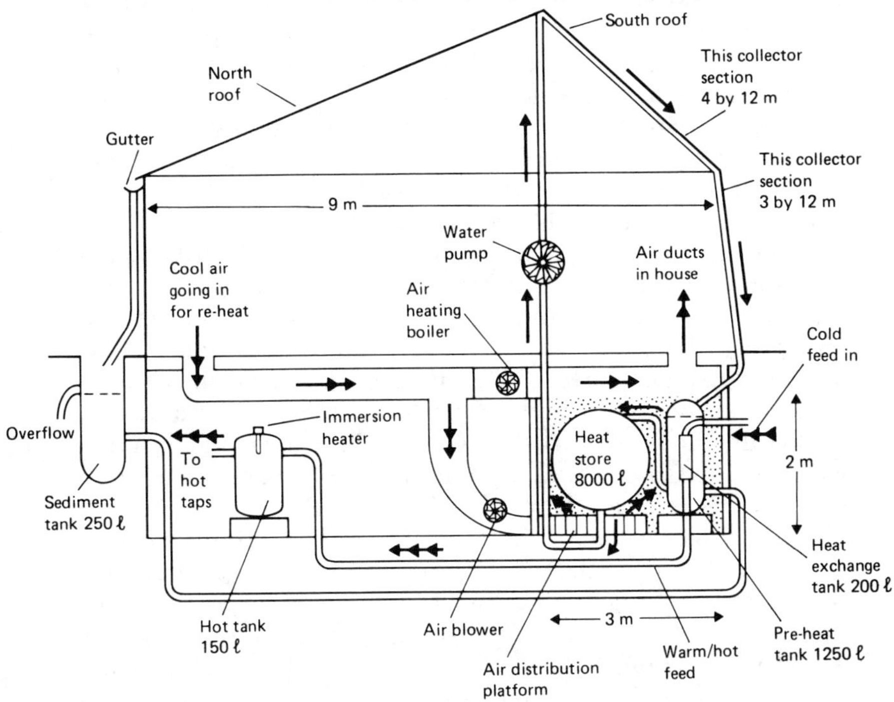

Fig. 102. Heat scheme 9. A single storey house with an equal area basement. Whilst one third of the basement is taken up with solar equipment, the resultant 6 days solar storage results in over 90% of the fuel bills being paid for by solar even in Washington D.C.

built pebble bin in which are buried two large tanks. The bin is 2 meters deep and covers a floor area of 3 by 8 meters – a total capacity of 48 cubic meters. The store tank takes up 8½ cubic meters and the pre-heat tank another 1½. As the pebbles (30% voids) store about 40% of the heat of a similar volume of water, the 38 cubic meters of them in the bin are equal to only 15 cubic meters of water, but together with the 10 cubic meters of actual water, the complete store is equivalent to 25 cubic meters of water, or 25,000 liters. In changing from the basic elements of Fig. 101 to a specific scheme in Fig. 102, we have added the heat exchanger in the pre-heat tank, and the hot tank with immersion heater. Yet another tank (a well painted and protected oil drum suffices) has also been added to act as a combined sediment trap and water level control for the complete water circuit. This later tank's function is to collect water from the North roof, to maintain the store tank *full*, and the pre-heat tank *90% full.* You will also note that the air blower and the hot air blower in the boiler are now two separate devices. One last important addition is the air distribution platform on which the store tank, and most of the pebbles rest. This is a good, cheap, simple way of distributing the concentrated air from the blower so that it is spread over most of the 3 by 8 meter floor area of the bin. Fig. 103 shows the arrangement. The idea is to build a platform from concrete blocks, setting the blocks so that the two holes in each are in alignment with the next block thus forming two air tunnels along each row of nine blocks, 24 tunnels in all. With 9 rows (each 12 blocks) *set 6 mm. apart,* the result is a well-distributed flow of air coming up over the whole platform. The blower supplying the air is fed straight onto four curved metal plates that push most of the air to left and right of center to enter two 200 mm. by 350 mm. tunnels each 2 meters long. The platform lies mainly under the 8000 liter tank (1.3 meters in diameter and 5.5 meters long) which is set to one side on the bin.

The bin itself is made by blocking off a 3-meter by 8-meter corner of the basement using similar blocks to those used in making the platform, and painting the inside with several coats

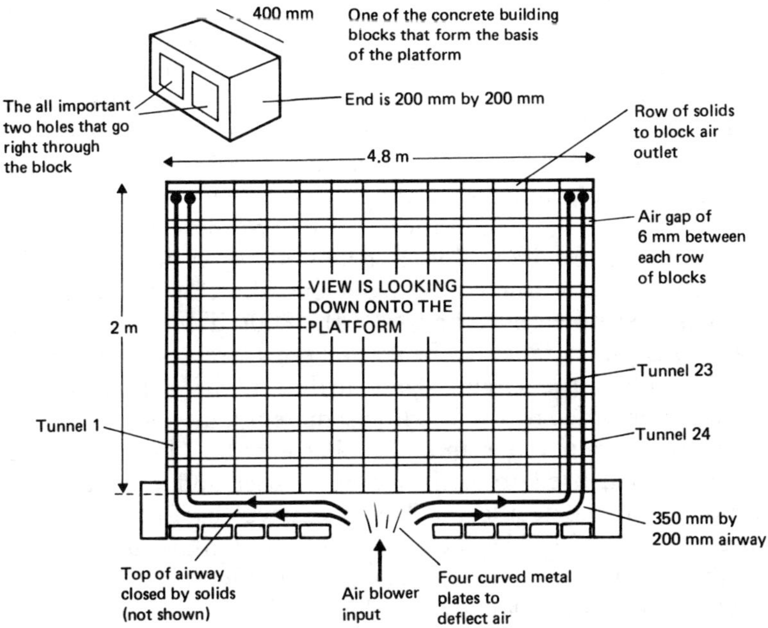

Fig. 103. The concentrated output of the air blower is distributed over a large area at the bottom of the pebble bin by blowing it into this 200 mm. (8 in.) thick platform so that it oozes out of the 8 slots each 4.8 m. (16½ ft) long.

of cement mortar mix to make it air and watertight. Heat insulation is added by first filling the airways of the blocks with insulation and then by putting a 75 mm. layer of insulation all round the inside of the bin. Finally, wooden protection is added to prevent the insulation from being crushed by the heat storage equipment yet to go inside.

Look back now to Fig. 102 as we work through the operation of the system. Cool water is pumped up to the top center of the roof (6 meters in from each end) where it is distributed across the whole 12 meters' width of the collector and allowed to trickle down its 7 meters to the bottom. (As you see, this is some collector – 84 square meters of it for a house area of 9 x 12 or 108 square rule of thumb.) The warm water is collected at the bottom center and piped straight into the 1250 liter tank where it not only heats the surrounding pebbles, but more important, transfers some of its heat to its internal 200 liter tank. This 200-liter tank is a third type of heat exchanger doing the same job as the finned tank and coiled tube types already discussed. It is worth considering whether the use of one of these more compact heat exchangers might allow a reduction in the size of the 1250 liter tank. The warm water, from one third up the 1250 liter tank, is next taken into the top of the 8000 liter store tank, and the cooler water taken from its bottom to return to the pump and complete the circuit. One further addition to this water circuit is provided by the 250 liter sediment tank and its function is to collect rain water and allow it to settle. The sedimentless water is then used to keep the 8000 liter and the 1250 liter tanks at their correct levels. This is achieved by first arranging it so that the 90% full level of the 1250 liter tank is *above the top* of the 8000 liter tank, and that the *overflow level* of the 250 liter tank is at the *same level* as the 90% level.

There is also one other, but quite separate, water circuit (three arrowheads in Fig. 102) that supplies the domestic hot water. This is its operation: cold water from the rising main comes into the top of the heat exchanger 200 liter tank, and is warmed by the surrounding water; it then continues out at the bottom of the tank at a temperature between 38° and 60°C (70° in hot climates) as the warm/hot feed to the hot tank. The immersion heater in the hot tank will boost the water up to 50°C (if necessary) before it is used in the house. Those are all the water circuits.

The air circuits start with the cool air going in for reheat at the left of Fig. 102 (double arrowheads). If there is sufficient heat in the Solar store, the air is directed down to the air blower and the passage to the boiler is blocked. The blower then takes the air and pushes it into the air distribution platform. As it oozes out from the platform it permeates up through the pebbles, past the tanks, is reheated and emerges at between 38°

and 60°C. This hot air is then fed to the house by the usual system of ducting. In the other case, when there is insufficient heat in the Solar store, the air is prevented from going down to the air blower, and is instead taken into the boiler for re-heat and on across the top of the pebble bin to the air ducts in the house.

The results achieved by this system are nearly as good as one can expect from Solar. The house has consistently obtained at least 90% of its heat requirements from Solar since 1959, but as you see, it is a system with a large collector and a very large store. Our two rules of thumb suggest that for a house of 108 square meters living area the collector should be 54 meters and the store 5400 liters. With a collector of 84 meters and a store equivalent to 25,000 liters the excess is clear. But this was a house designed to get much better than a 50% to 75% payoff, and that it does. With its Solar store over four times the norm, one can readily accept that it has a 6-day carryover; and this makes the difference. As the house is in Washington, D.C. at Lat. 39°, it is at the better side of our band 3, and thus, one would expect 1½ days from the *normal* size store.

For those who find the size (particularly of the store) somewhat daunting, a scaled-down version could be considered, but if this is carried out one can hardly expect to obtain as good results. As you will appreciate later, the cost of getting a greater or lesser percentage of your heat bills paid is not necessarily proportionate to the expenditure on Solar equipment.

Together with the notes at the chapter end, this completes the heating part of scheme 9, but, perhaps to your surprise, the scheme as shown in Fig. 102. First, an extra outlet pipe is methods that can be used to add cooling.

In the simplest of the two, these are the only alterations to the scheme as shown in Fig. 65. First, an extra outlet pipe is taken from the 8000-liter tank, at a point two-thirds of the way up from the bottom, and made available for connection to the water pump. Second, a pipe is fitted along the full 12 meters of the *North* roof so that water pumped up to it will sprinkle out and run down the full length and breadth of the roof. The two new pipes can then either be connected to a second water pump, or, can share the existing one by using two, two-way changeover devices. The pump now has two functions. In the *winter,* it is lifting water up in the *daytime* to enable it to run down the *South* roof and get *hotter;* and in the *summer,* it is lifting water up at *nighttime* to enable it to run down the *North* roof and get *cooler.*

The summer cooling mode, operated mainly by night evaporation on low humidity evenings (preferably with a slight breeze) achieves cooling by stright evaporation, and, as a consequence, there is some loss of the circulating water. Check the level in the 250-liter tank during dry spells and fill the tank up as necessary. This is the operation of the night cooling. The warmest water in the 8000-liter tank (that's why it was taken from two-thirds up the tank) is pumped up to the apex of the roof and released from a series of holes in the pipe to form a continuous spray at the top of the roof. The water running down the roof cools itself, partially by contact with the cool roof, but mainly by night evaporation. As this cooled water is collected in the bottom gutter, it is fed into the 250-liter tank gradually cooling it. This in turn feeds into the 1250-liter tank and the 8000 liter cooling them both. The last two tanks also cool the pebbles, so storing the night 'coolth' – to be ready for daytime use. To recover the stored coolth during the day requires that the blower is used in the usual way, but this time it sends out cool air and receives warm. At night we are trying to push the temperature of the store down, and during the day use it is being pushed up again. The insulation around the bin, like everything else, works well in reverse, except possibly the 200-liter heat exchanger. If one succeeds in getting the bin below the incoming water temperature of 12°C, one is then cooling the water still further before paying for the immersion heater to heat it up again! This point is mentioned more out of academic interest than as a suggestion that the 200-liter tank be bypassed to solve the problem in the summer, as the difference is not great.

The second method of providing cheaper cooling is based on the practice of some utilities, in charging only half the normal rate for nighttime electricity. If your area is one of these, the idea is simply to fit a normal air conditioner to cool the pebbles *at night*. The cooling being fed into the bin at any convenient point in the ducting prior to the air blower. One is then cooling the bin of pebbles at night with half-price electricity, and using the stored coolth during the day. There is really no reason why both schemes cannot be used to cool the bin – the very cheap roof cooling on low humidity, breezy nights, and the half-price air conditioner cooling on still balmy nights.

The following notes apply to all variations of heating and cooling schemes.

1. The water distribution pipe going along the top of the collector has to be copper – plastic would melt!
2. The fresh water pipes have to be copper as they are under pressure.
3. All other pipes can be of plastic (polyethylene, etc.).
4. The air going for re-heat is better if passed through a filter.
5. The water pump(s) should be ¼ to ½ h.p.
6. The pebble bin is filled to within 300 mm. (12 in.) of the underside of the floor joists.
7. The spaces between the floor joists in the area over the bin should be filled with heat insulation to prevent unwanted heat from getting into the house.

A reduced complexity, and reduced scale, version of the Thomason system is in use at the B R A D commune in Wales, U.K. They use it only to produce hot water at the moment – for all eighteen of them! The simpler version of the collector was constructed with the aluminum panels straight onto the rafters, and single glazed with glass reinforced plastic sheeting. Fig. 104 shows the general detail; and as can soon be appreciated, one could extend the B R A D version to provide warmth.

During the life of the above system, the people at B R A D have carried out some experiments of their own. The point they make is that with the relatively small volume of water that they have in their closed loop, they can put in certain additives and observe the effect. These are the points they have made.

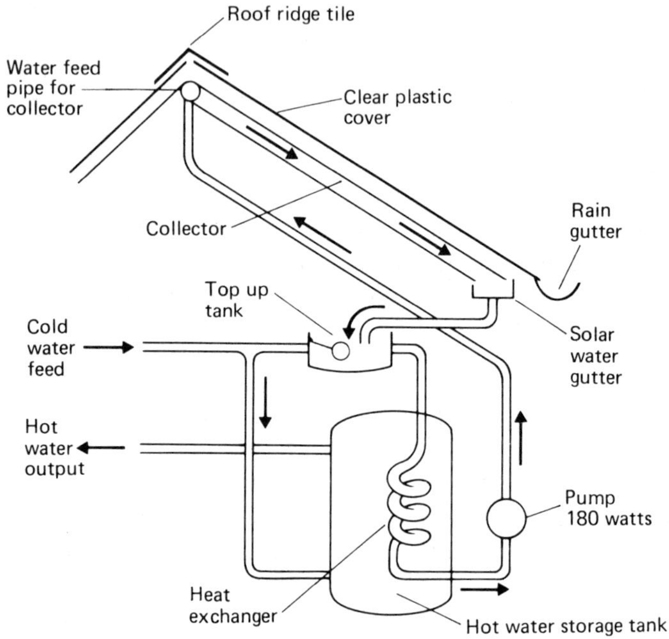

Fig. 104. This simplified 'Thomason' provides hot water for 18 residents of a Welsh commune.

1. The system runs well as there is not a constant stream of new salts arriving in the water to corrode or furr it up.

2. An additive such as Indian Ink can be tried in the closed water loop to increase the Solar absorption.

3. An algae combating agent such as copper sulphate can be added to the water loop.

4. A quantity of detergent can be added to aid the 'wetting' action of the water thus giving a better spread across the aluminum, and consequently better Solar pickup.

Patents. This chapter is the most appropriate time to mention this matter, as not only was Harry Thomason very much involved with patents before moving into Solar, but also because he holds more Solar related patents than any other person we know.

Many of the ideas featured in this book are patented. The purchase of this book neither grants, nor implies, any patent right to you. If the patent is filed and granted in your country, you certainly cannot use the idea in any item you produce for sale without obtaining the agreement of the patent holder. If you make one for your own use, you generally have no problems outside the U.S. This is true for many reasons; the fee is too small to be worth collecting; the holder of the patent would have to be certain you were using exactly his idea; your country may permit 'one-offs' to be made for self use; the patent may not be operative in your country.

Having given the ground rules, here is one exception! Harry Thomason has arranged that detailed house plans, with written knowhow, are available from an associated U.S. company. They are also empowered to issue licenses to build at $20 each. As the total outlay for the information, plans, and the license represent less than 1% of what you are liable to spend on your Solar installation, it is a very worthwhile investment. It is a small but well-earned recompense for this very able pioneer. Details are given in the Directory.

CHAPTER 14

A Backyard Solar Furnace

Here we see a new presentation of Solar Energy hardware that will undoubtedly be the forerunner of a different breed of products. A privately funded research group, International Solarthermics Corporation, led by John Keyes, have come up with a new concept rather than a new invention. The concept is simply to put together what they consider are all the best ideas and arrange them as a complete factory produced package. This single package is not part of a system, but a complete system ready to 'plug into the house'.

That Keyes and his team have broken little new technical ground does not belittle their efforts, much effort and brain power is required to select various parts of a system and match them together in such a way that a factory can repeat the design and produce an economical and reliable product. This is the sort of slogging effort that Henry Ford referred to when he said, 'invention is 5% inspiration and 95% hard work'.

The advantages of a mass-produced package like this are: that the customer knows pretty accurately what he is paying for, and pretty well what he is going to get from it in the way of heat. Such a package could (and in Keyes's case does) have a guaranteed performance, both when delivered and after a year's use. Furthermore, this guarantee covering the performance of the whole system is obviously better than a guarantee of a component part of a system. This way, if there is any trouble, there can be no argument as to whose part is letting the system down.

The disadvantages of such a system are: that there is less

opportunity to reduce costs by building it yourself (some merchants do supply it in kit form which helps); the package takes up room in the garden; and, it is only earning its keep in the winter as it does not produce hot water in the summer, or indeed at any other time. This last objection should be overcome before long as there is no fundamental reason why such a concept should not provide hot water.

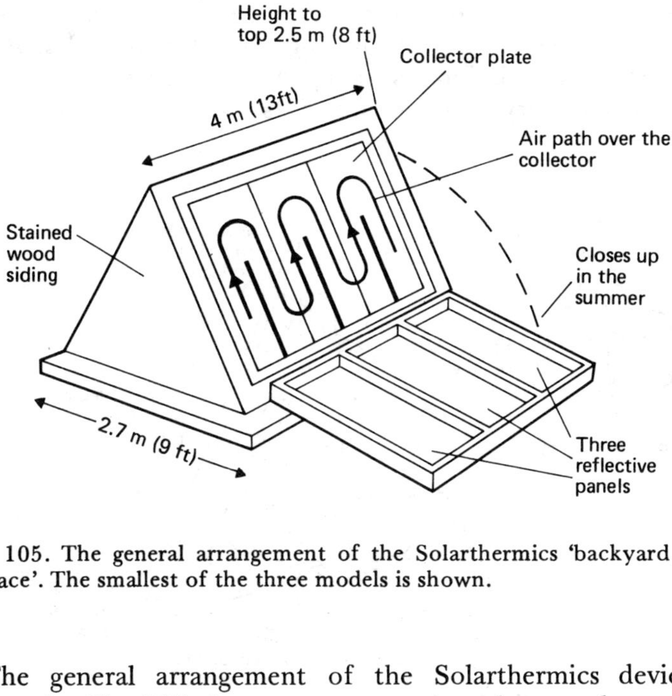

Fig. 105. The general arrangement of the Solarthermics 'backyard Solar furnace'. The smallest of the three models is shown.

The general arrangement of the Solarthermics device is shown in Fig. 105, here you see a triangular wooden garden shed type structure nearly 2.5 meters (8 ft.) high with one glass side facing the Sun. Inside the shed type furnace, warm air is collected behind glass and blown round into a pebble store contained at its center. A second blower circulates warm air from the furnace's store to the house via a connecting duct, the cooler 'used' air being returned by a second duct to the furnace.

The detailed construction of one corner is shown in Fig. 106. This shows a double glass cover over an unusual black surface designed to heat up the air. The unusual surface is made up of what appears to be blackened beer cans. They are actually aluminum cylinders (as are some beer cans) blackened inside and out, with their closed end resting on the collector plate, and their open end facing the Sun. The collector plate is a flat sheet of metal, and we see no reason why aluminum cans should not be used for the cylinders. The cans act to absorb more of the Sun's rays by making them hit the sides of the can several times before escaping. At each hit more energy is absorbed and very little actually gets out. See top left of Fig. 106.

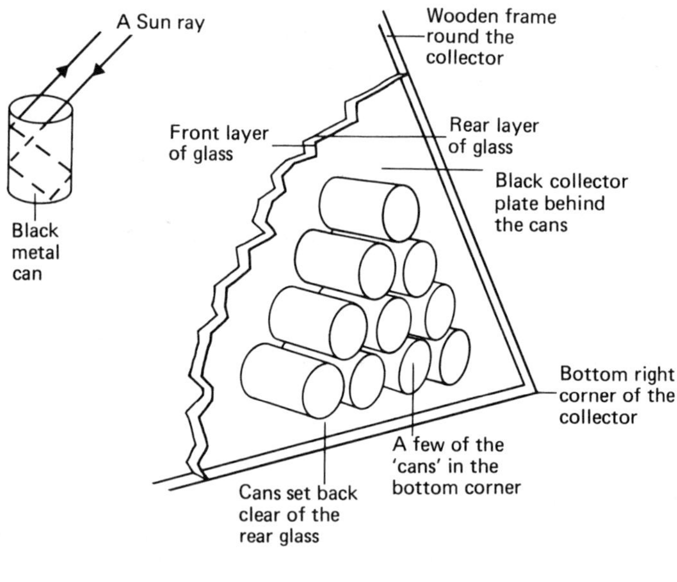

Fig. 106. A cut away section of the 'can' collector. The top left inset shows how a Sun ray has several chances of being totally absorbed as it 'bounces' round a can trying to get out. This is a version of a 'vertical plate' collector.

Having intercepted as much Sun as possible, to get the cans as hot as possible, a controlled air flow is now created across the cans. The flow is made to traverse a serpent like path with the aid of three baffles, so that the air passes over a greater number of cans. In addition, by keeping the air flow going across the cans at a slight pressure, the air now takes on a special 'washing' like action. Instead of merely passing straight over the cans it eddies round literally washing the heat out of them. It is detail like this that makes the difference between mediocre and good design.

Referring now to Fig. 107, we can see the internal workings of the furnace behind the 'can collector' we have just discussed. The hot air leaves the collector on the right of Fig. 105, thus appearing at point 1 on Fig. 107. To enter the furnace's store area, the hot air is first taken through a heat trap before emerg-

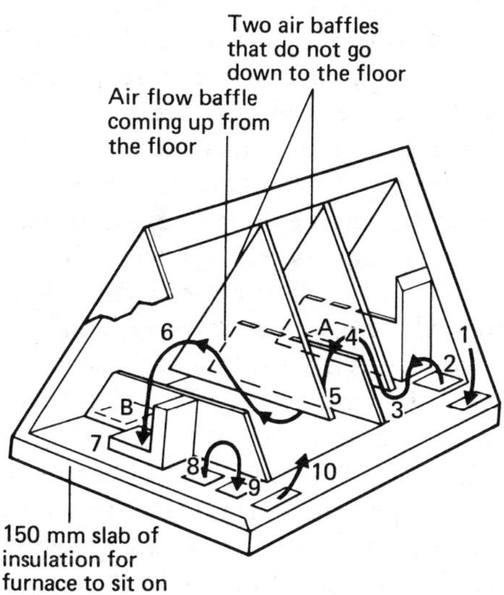

Fig. 107. The inside view of the furnace shows the air flow after it has come from the collector at 1 to when it goes back to the collector at 10. A shows where the other air flow path starts out for the house and where it comes back at B.

ing at 2. The idea of the heat trap is to prevent the stored heat from escaping back to the collector at times of no Solar radiation, night, etc. As the air continues over and under the three internal baffles, 3, 4 and 5, it is actually passing *through* the pebble store, which fills most of the center of the furnace. In so doing it gradually gives up its heat before arriving at 7. A further trap is installed after 9 before the cool air arrives back at the start of the collector at 10. The fan that provides this first air flow is installed at point 8.

The second air flow circuit is the one that delivers the hot air to the house. This air is blown by a second fan and goes out at A to arrive back at B. As can be seen this is in effect a *reverse* flow across the pebble store. If both fans were going at the same time, the heat would tend to come off of the collector at 2 and go straight out at A, rather than attempting two opposite air flows through the pebble store area.

The general construction of the furnace is made up from laminated plywood forming a sandwich around polystyrene foam. Reflective foil is placed on the inside, plywood facing the pebbles. The sandwich walls are 100 mm. (4 in.) and the back 150 mm. (6 in.) thick, the whole furnace sits on a further 150 mm. slab of insulation. Great care has been taken with the general engineering to ensure long life; galvanized metal parts, sealed bearings for the motors, etc. Quite sophisticated electronic control is provided for all functions including a protection device that brings the blower motors on gradually. It appears that if a hot collector has a good blast of cool air applied to it it is likely to shatter the inside glass. The usual two comparative temperature sensing devices are also included to ensure that whenever the collector is above the temperature of the store the blower is activated to put the extra heat into store.

The supply of warm air to the house from the furance is set by the house thermostat. When heat is required, first call is made to the Solar furnace, if its store of heat is used up and it is not producing any more from the collector, the command is transferred to the house furnace. A typical sort of operation that we have already discussed.

The performance of this equipment is very much a feature of the concept, and not only is the performance stated but it is given as a guaranteed minimum. As the performance must alter in different areas in the country, the location is stated at the same time as the performance. To state a typically good central position we will give the minimum performance of the Model 96 (that's the one in the Figures) at Denver, Colorado. Lat. 40°.

The specified performance is given separately for each of 5 2-monthly periods from mid-October to mid-June. At each date the rating says, 'for this cloudless strength radiation of . . . the furnace will give . . . much output'. It also states that, 'if you leave the heat in store for 3 days at . . . outside temperature, it will give you back . . . percentage of what you stored'. That is good basic information; it tells you how much it collects and how well it holds it. But as the data only covered cloudless skies we have also added typical condition figures.

This is the data for operation at Denver.

Heat arriving on one cloudless day (horizontal plane)		Amount that Model 96 (8.2 sq. meters) collects on that day	Likely average daily collection at 40° lat.
October	4.6 kw/hr meter	42 kw/hr	28 kw/hr
December	2.5	35	17
February	4.1	42	24
April	7.0	40	27
June	8.4	30	?

The figures show two things. One, that due to the tilt of collector, etc., the amount collected does not follow the amount arriving. Two, that the action of the reflector and the angle of the collector makes the Model 96 pick up an 'impossible' amount of heat sometimes.

If you require further information write:

International Solarthermics Corporation
P.O. Box 397
Nederland, Colorado, 80466, U.S.A.

Remember, I.S.C. neither manufacture nor sell products, they design in order to license other people, and will be only too pleased to tell you who they are.

CHAPTER 15

Wind and Water Power Schemes

A recent calculation showed that if a windmill were put atop of each electricity pylon they would produce twice the power of all the power stations that feed the pylons! It is one of those schemes that never is implemented, but does show the amount of energy that is available.

As these two other forms of Solar Energy wind and running water both produce rotary motion, we will consider them as essentially the same. Actually the water mill produces rotation around a horizontal axis, but the windmill can produce horizontal or vertical rotation, and, of course, nearly everybody has a supply of wind power. From now on we are only referring to wind power.

In the chapter we will look at the amount of power there is in the wind; the various methods of collecting it; and how best to use it when it has been collected.

Wind is cool air moving in to replace hot air that has risen up from the ground as a result of the Sun's heat. The speed of that wind is fastest as it gets clear of the Earth, because lower down it is slowed as it rubs over the ground surface. We soon learn then, that to capture the full power available we have to get up clear of the ground. Fig. 108 expands the point and adds the general considerations to bear in mind when choosing the site for your windmill. They include:

1. Fixed to an out building or tower. (The roof of your house is not too good a choice as windmills often set up an annoying hum, but if you can shut it down at night you may find it satisfactory.)

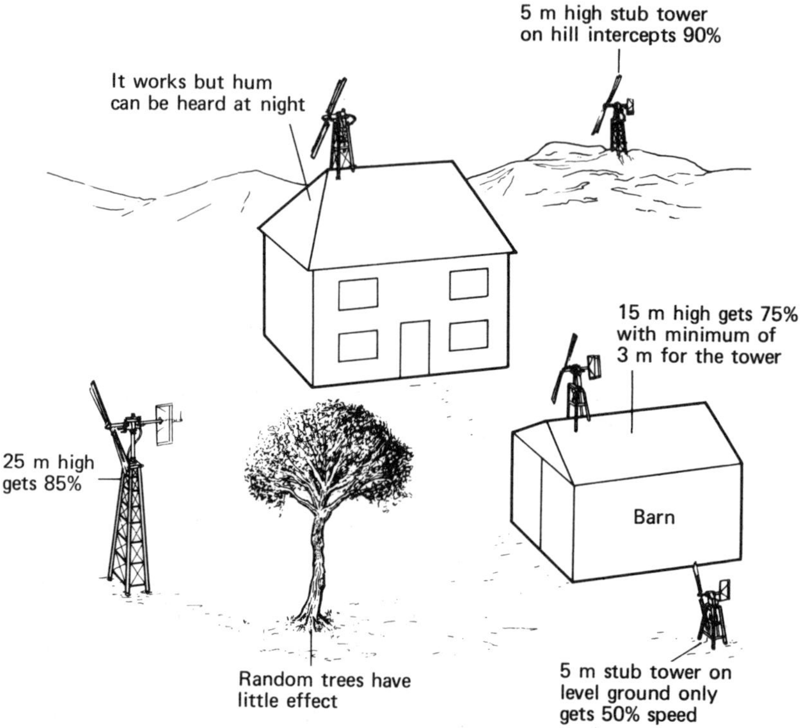

Fig. 108. Where to site your windmill. 'Full wind speed' occurs about 50 m. (160 ft) above level ground, we show the percentage of that speed at lower heights. The loss is greater than it appears as a 50% speed loss (2 to 1) is an 8 to 1 power loss (it goes down by the cube).

2. On top of a small hill and 5 meters (16 ft.) clear of the ground; but remember you do have to get the power to where you wish to use it with minimal losses.
3. Clear of ground obstructions such as buildings or groups of trees (oddment trees and small buildings do no harm).
4. On a pylon or stub tower – preferably 15 or more meters high.

Throughout the world there are very few areas where there is not sufficient wind to be worth collecting. Europe is well served by Westerly winds that cover much of the land mass in the winter, and the more local winds such as the Mistral at other times. Australia has its South East Trade winds in the Northern Territories, plus Brickfielder and Southerly Busters lower down. New Zealand has its Nor'wester, South Africa its Berg winds, Canada its Westerlies, and generally all over the U.S. there is a steady wind harvest waiting to be tapped. The Doldrum Calms are indeed limited to a very small part of the Earth's surface, and to be precise they are over water.

Winds are consistent from year to year and in general more predictable than other forms of variable energy. Moreover, their pattern over the year is very suited to our needs as they are nearly always strongest in the winter. This particular form of Solar Energy is an ideal complement to the radiant form by supplying a fine boost to that weakening winter supply.

Fig. 109 shows a wind energy graph similar to the Solar radiation graphs at the end of Chapter 8. The lines in the wind graph show the amount of energy from the various wind speeds in each month of the year. The bottom energy line is of no value because windmills rarely start at these low wind speeds, the top line also contributes little as many windmills have been taken out of action at these speeds for fear of destroying themselves. From the graph you can see that in the month of September there is the usual small amount of energy due to the 6 to 11 km/hr winds (4 to 7 m.p.h.), but there is enough power in the 13 to 19 km/hr winds (8 to 12 m.p.h.) to produce a total of 15 kw/hrs in that month from a swept area of one meter; about ½ kw/hr per day. The graph also shows enough energy in the 20 to 29 km/hr winds (13 to 18 m.p.h.) to produce a further 30 kw/hr in the month, and so on until the total for the months for all winds is shown. The exact figure is 67 kw/hrs less the 4 kw/hr for the lowest speed wind that was not useable. There were no times that the wind was above 51 km/hr to cause the mill to be closed down. Note carefully that the 63 kw/hr (about 2 kw/hr per day) is the power *in the wind,* and is the

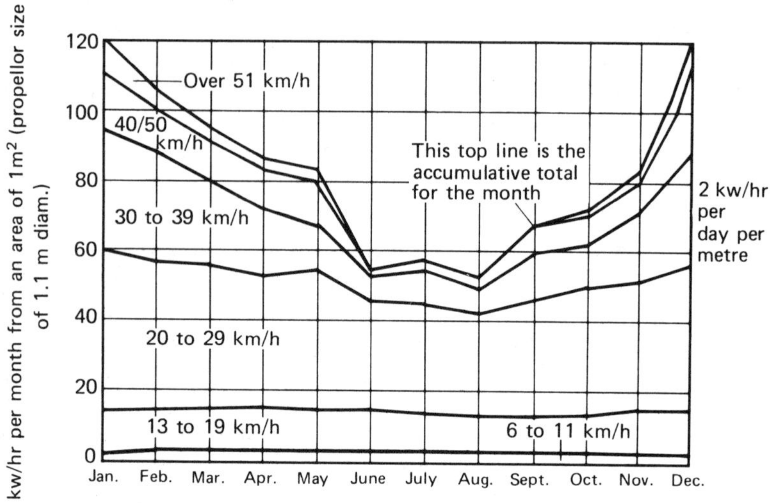

Fig. 109. The amount of energy in the wind each month of the year at various wind speeds. The total energy arriving each month, as shown by the top line, is the important one.

amount intercepted by one square meter of swept area.

By totalling up the 107 kw/hrs for January, the 97 in February, etc., the year's total can be seen to be a little over 900 kw/hrs. But as with Solar radiation there are losses to allow for as one attempts to gather the energy, and at the best a bare 60% can be intercepted by any wind device. Furthermore, even if the interceptor is as good as a high-speed propellor, it will only have an efficiency of 70%. Now our 900 kw/hr goes quickly down to 540 kw/hr and then to 380 kw/hr as it appears as the equivalent mechanical rotation at the windmill shaft. While these month-by-month wind energy data should be available from the 'Weather Office', TV station or local airfield, it is actually less important to have them than the Solar radiation data. The pattern of wind energy supply for the various months of the year will often suit your demand (coming more in the winter) and so requires less careful matching of supply and demand. Another method of assessing your local wind energy income is to consult a manufacturer of wind equipment. Manufacturers do have such information, but insist on a month-by-month statement as they tend to offer it as 'yearly total of kw/hrs collected for each 1 kw. of installed generation'. The limitation of this form of information is that much of the energy could arrive during only 2 or 3 months of the year, and thus either have to be partially wasted, or, require effective long-term storage.

As a first assessment of whether you have sufficient wind power in your area, you should establish that you have winds of over 16 km/hr (10 m.p.h.) on at least 2 or 3 days each week. This data is always available from radio, TV or newspaper reports. Winds of 10 km/hr (6 m.p.h.) are the starting point for collecting wind power and are often referred to as the 'cut in speed' of a system, there is also some energy in the 'prevailing winds' between 10 and 20 km/hr (to and 12 m.p.h.), but the main collection is from the 'energy winds' between 20 and 40 km/hr (12 and 25 m.p.h.). These last two groups are referred to as the 'rated speed' of a system. At winds above approximately 50 km/hr (30 m.p.h.), most systems reach their 'furling speed', which means they are put out of action until the wind drops. Better systems are those that go into a constant speed mode in high winds, as these continue to produce full power.

If the detail of the various wind speeds in your area are available, each one can be converted to kilowatts collected for a certain propellor size by using the Table in Fig. 110 (as most wind data records will be in Imperial units we have not included the metric), each can then be added together to arrive at a total for the month as in Fig. 109. If only the average wind for each month is available in your area, Fig. 111 shows how to convert that information into the amount of energy you can expect in each month. The kw/hrs stated are the actual amount you will get after allowing for the normal 75% efficiency of a generator. In our prior example this will bring the 380 kw/hr down to 285 kw/hr from that 1 square meter in a year.

Prop. diam. in feet.	Wind velocity in m.p.h. 5	10	15	20	25	30
4	2	19	64	150	300	520
6	5	*42*	140	340	660	1150
8	10	75	260	610	1180	2020
10	15	120	400	950	1840	3180
12	21	170	540	1360	2660	4600
16	40	300	1040	2440	4740	8150

Fig. 110. The power produced by various sizes of propellor in various wind speeds. This is energy of rotation at the shaft, if it were used to drive a generator only about 75% of it would appear as electrical energy. Note, how twice the wind speed gives 8 times the power, or, twice the propellor size gives 4 times the power.

Size of gen. output in watts.	Average monthly wind speed in m.p.h. 6	8	10	12	14	16
250	6	12	18	24	29	32
500	12	24	*35*	46	55	62
1000	22	45	65	86	104	120
2000	40	80	120	160	200	235
4000	75	150	230	310	390	460

Fig. 111. The amount of energy given out, by various sizes of generators, in Kilowatt hours per month for different average wind speeds in a month.

The link that connects the previous two Tables is as follows. The first Table shows that a 6-foot propellor in a steady 10 m.p.h. wind will produce mechanical energy equivalent to 42 kw/hrs in a month, and if this is used to rotate a generator it will be converted to 32 kw/hrs (75% efficiency) of electrical energy. As the second table shows that 35 kw/hrs is produced in a month (nearest to 32 kw/hrs) from 10 m.p.h. winds and a 500-watt generator, it follows that a 6-foot propellor is required to drive that generator. Higher average wind speeds in an area mean that a smaller propellor will suffice for the same size generator.

The average annual wind speed over much of the U.S. is 16 km/hr (10 m.p.h.) with a tendency to be somewhat higher nearer the coast. In countries such as the U.K., which are nearly all coast, their annual average is more like 20 km/hrs (12 m.p.h.)

Having obtained a general feel of the wind and its capabilities, the choice of windmill can now be considered as the previous discussion only related to the amount one could collect.

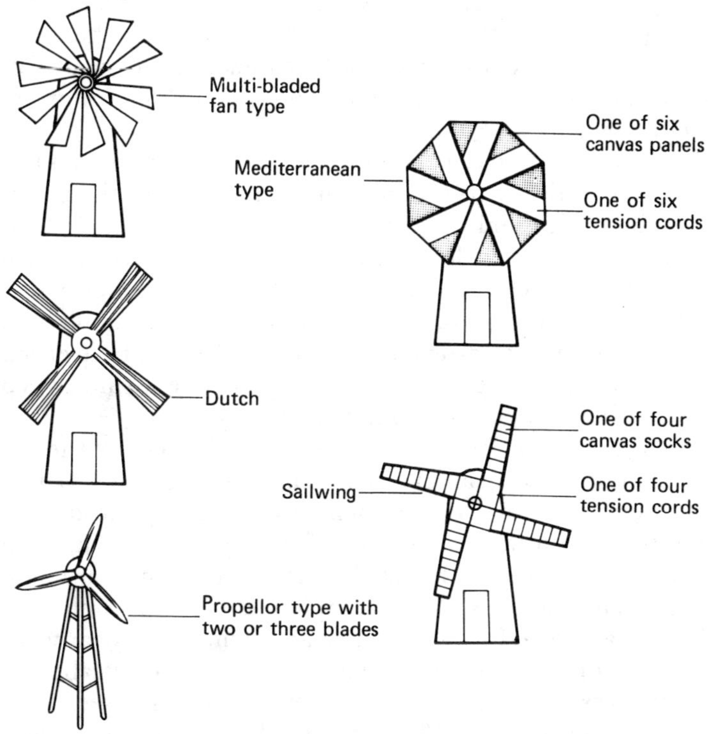

Fig. 112. The five main types of horizontal axis windmills. All these devices have to be kept facing the wind to gather energy.

Windmills fall into two main groups, the well-known Dutch type which rotate around a horizontal axis, and the newly re-discovered crop of vertical axis windmills. Fig. 112 shows the main horizontal types. At the top we see the 'multibladed fan' type of English windmill. A mill that starts easily with low wind speeds and rotates relatively slowly (100 to 200 r.p.m.), but whose efficiency is often below 50%. Next, the 'Mediterranean' type with its 6 or 4 paddles made out of canvas triangles held taut by a cord going back to the next arm. Then comes the 'Dutch' windmill, famous for its role of pumping water out of the Lowlands of Holland. Following it is the 'Sailwing' which is really a hybrid of the previous two types, it is formed from a sort of canvas sock slid over a main spar and held taut by a tension cord running nearly parallel to the spar. Finally, the 'propellor' type which usually has two blades, a third sometimes being added to aid starting.

Five basic types that gradually move from a maximum full-bladed circle down to a minimum of two. When the blades are at maximum, the mill is the easiest to start at low wind speeds, but it rotates at the slowest speed and has the lowest efficiency of about 50%. The two-bladed variety at the other end is the hardest to start, but rotates at the highest speed (500 r.p.m.), and has the highest efficiency of nearly 70%. (These efficiency figures are in very general terms and assume a similar type of construction, which is often not the case.)

We move now from the fairly conventional to the unconventional, the vertical axis windmills of Fig. 113. The main limitation with this type is their low efficiency; at any instant when they are rotating, one half of the device is moving *towards* the wind. This means that as that half is not contributing to the energy being collected, the complete device can only be 50% efficient at best. In fact, as the advancing half nearly always detracts from the collecting half, the efficiency is about 40%. Nevertheless, vertical types have their virtues and uses.

At the top of Fig. 113 is the popular 'Savonious Rotor', the delightfully simple-to-make device referred to at the end of Chapter 4. Lower in the Figure is the 'Darrieus Hoop', another

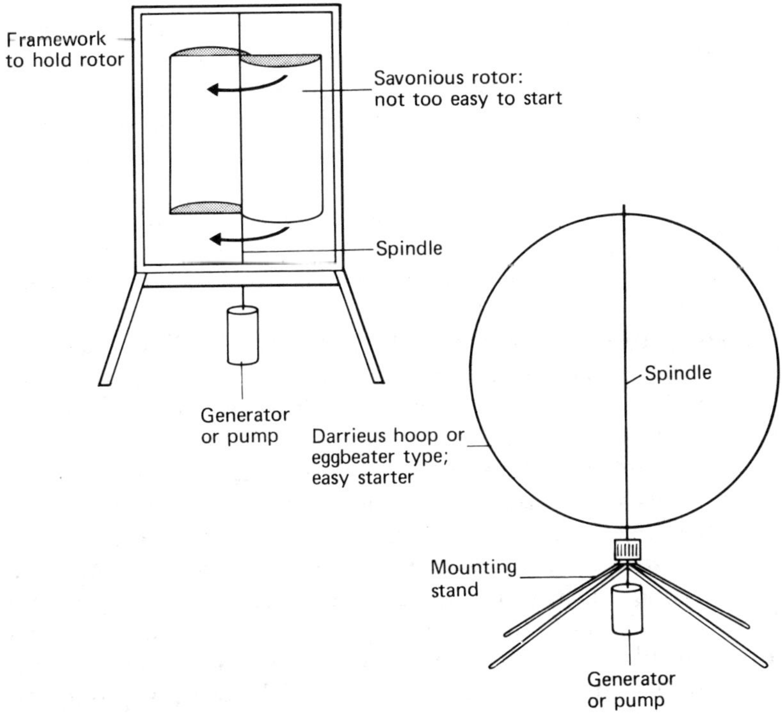

Fig. 113. Vertical axis windmills operate from winds coming in any direction. They also allow the rotating shaft to be extended down to the ground, but they are not very efficient. One variant puts the rotor inside the hoop.

simple device very much as it appears, although the actual strip forming the hoop is of aerofoil section. The hoop is a device that will start in very low winds, and whilst it appears largish at 5 meters (16 ft.) in diameter to produce a modest 1 kw. output (suitable for a home) from a 15 m.p.h. wind, the configuration is somewhat elegant and attractive to behold.

A story is told of a large U.S. aerospace Corporation who recently re-discovered the hoop! Upon starting on a programme of research into new windmill types, they made the classic mistake of not checking the 'prior art', and after eventually coming up with their 'new' invention, were somewhat dismayed to find that it had been patented by the French inventor Darrieus in the early 1900s, consequently it was available for all the world to see and use.

This last design is about the lowest efficiency device of all the seven we have discussed, but as you may consider that a hoop of 5 meters diameter doesn't really appear to take up nearly 20 square meters, it may well be acceptable.

Distinct from normal windmills, the vertical mills operate just as well whichever way the wind blows at them, and do not have to swing into wind to be effective. Additionally, the power output, as rotational energy, can be brought down to earth very easily; a mere extension of the rotating shaft is all that is required. This means that the pump or generator can be at the bottom of the tower, so reducing weight at the top, and keeping the generator out of the weather. A further complication of the normal windmill is also avoided as the electrical wires carrying the power down can now be wired direct to the ground. When a normal windmill moves round to face the wind (complete with its generator), its wires have to be fed via electrical 'slip rings' to avoid wrapping them round and round the tower. These rings are an extra expense and a potential source of trouble that the verticals avoid. We will now consider the best use of this rotational energy.

The speed of rotation of all of the windmills we have discussed is never fast enough for normal generators, although some lower speed types are coming along. Our mills rotate at 100 to 200 r.p.m. for the multi-bladed fan types (plus the vertical types) and up to 500 r.p.m. for the two-bladed propellor. Most generators (which produce D.C.) and alternators (which produce A.C.) start to function at much higher speeds. An auto-generator will start producing quite well at 700 r.p.m. but is really designed to operate over the range of 1500 to 3000 r.p.m. This speed discrepancy requires the use of gears between the wind rotated shaft and the generator. Ratios of from 3:1 to 10:1 times are required for propellor types and from 8:1 to 30:1 times for the fan types. As this gearing unfortunately hinders the start of some of the faster rotating devices, techniques have had to be developed to remove the electrical and/or the mechanical load until rotation has commenced.

Having obtained our power, we assume that some form of storage will have to be included before it can be made good use of. There are four possible methods.

1. Water is pumped up to a higher level for later release.
2. Air is compressed for later release.
3. Electricity is generated and stored in batteries.
4. Electricity is generated and used to heat water.

A fifth idea is now gaining some favor, to drive (rotate) a heat pump or cooling compressor, but as there are still some problems due to the intermittent nature of the supply, it is not included.

The first two proposals require no further elaboration as they have been discussed in Chapter 5, and are schemes that either suit your situation or are totally out. Item 3 is a possible for everybody, and over the years is the one most used by windmill proponents. This association between windmills and electricity came about when windmills were used to supply light for outlying farms and homesteads where no main supply of electricity was available. The association lives on, but with two provisos we have to disagree with this popular concept. Providing that mains electricity is available, and, providing that electricity is used to heat water (even to complete the heating of Solar heated water), then it is uneconomic to use wind generated electricity to provide light. It should, instead, be used to heat water. The reason is quite simple; the cost of the storage batteries alone (which have a limited life) is the same, or more than the value of the electricity that they save during their life. Put another way; as each X pence of electricity is put in and out of a battery, it causes about the same X pence worth of life reduction to the battery. A way to store the energy of the electricity – assuming one is not using batteries – is to use it to heat and store water! Saving a unit of heat electricity is the same as saving a unit of light electricity. The only equipment required to enable this more efficient use of windmill electricity is one additional immersion heater added to the domestic hot water tank. The heater has to be of a special low voltage to suit the generator or alternator, and might be as low as 12 volts. The generator or alternator is connected directly to the heater and 100% of the electrical energy will appear as heat in the water – quite novel that there is no loss!

As a final suggestion, one could consider using a combination of two of the four schemes; for example, if the electrical power from a generator was not required, the wind energy could be diverted to pumping water.

Some form of tower will be required to mount the wind generation equipment on, this can either be a stub tower or a full size pylon, but always the watchword is *strength.* The wind speed experienced by a tower can often be five times the mean operating speed – and that gives stresses 25 times more than normal. Keep the design open in construction to reduce wind pressure, and make extensive use of guy ropes (wires). Full-size towers are a considerable part of the total system cost. One innovator reduced his tower costs by designing a circular base made out of concrete blocks, the base doubling as a water store for his Solar collector. The arrangement gave him great mass at the base of the tower, good strength and excellent stability; altogether a good example of the double use of one's investment.

An example of a typical home-made propellor driven wind generator is shown in Fig. 114. Most of the basics are shown but the generator (as in other figures) is shown as a direct drive from the propellor. This is only possible if one can find slow-running truck generators of the 12-pole variety, if not, the more usual gears will have to be added. Gears have an advantage in that the stress of the propellor is no longer applied directly to the generator bearings, which were not designed for that sort of load. If general attention is given to strength, resistance to weather and vibration, the unit will give valiant service. Bicycle

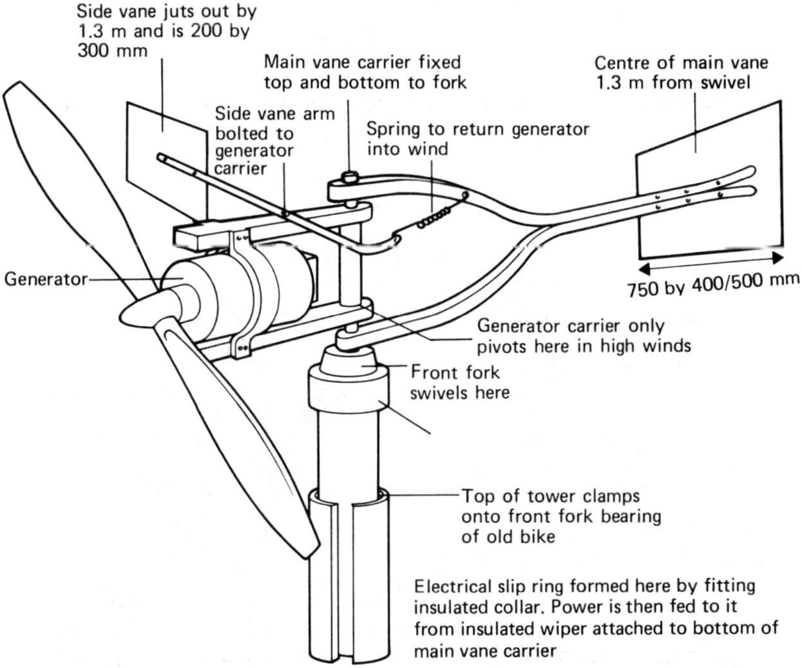

Fig. 114. A simple small windmill shown in its normal running configuration. The main vane is always down wind but the side vane will carry the propellor and generator to side if the wind is too high.

parts are quite useful to construct these small mills, while auto transmissions parts serve well in the construction of larger systems. The Directory leads to suppliers of other and more detailed plans.

All propellor windmill devices (and most others) have to include a safety device to reduce or stop their operation in winds above 50 km/hr. There are more designs for these devices than there are windmill manufacturers, but some reliable method of furling, feathering, or otherwise shutting down the windmill has to be fitted. The devices come in two groups.

1. A centrifugal force device that alters the angle of the propellor as the speed of rotation increases, thus tending to alter the mode to constant speed operation.
2. A turning device that gets the propellor out of its normal attitude of facing into the wind. These are either the 'flip round to one side' device as in our Fig. 114, or a similar design where the propellor flips over the top.

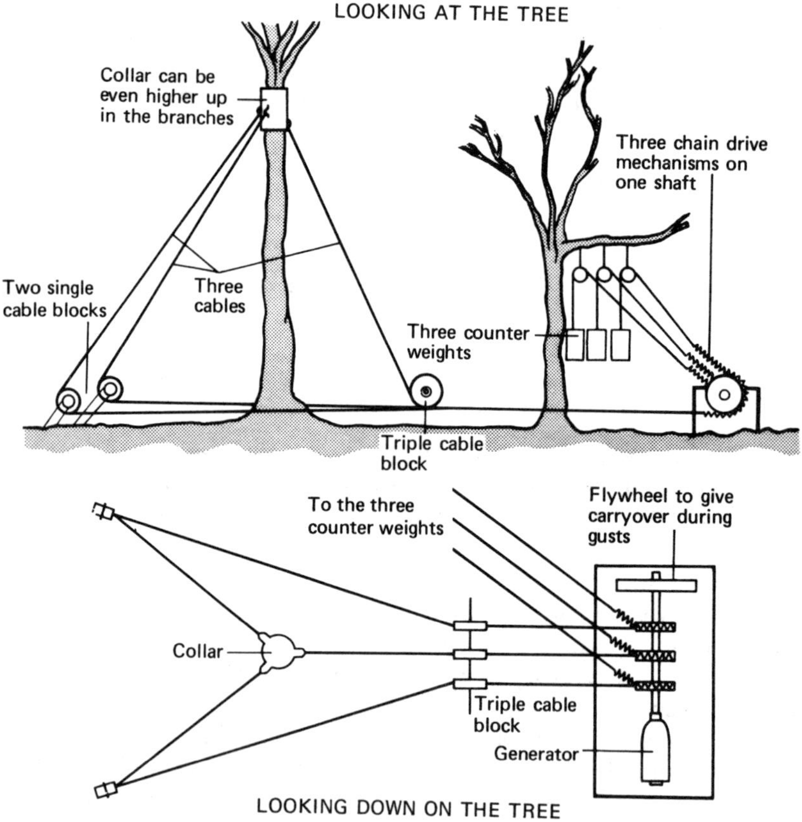

Fig. 115. The tree pump. Whichever way the tree is bent, as it intercepts the wind, it moves one of the three cables. The cable pulls its own chain driven wheel to keep the generator turning.

An even more sophisticated design gradually turns the propellor more and more out of wind as its speed increases.

Experience gained over many years suggests that careful attention should be paid to safety with windmills. The rugged, and of necessity, exposed life that they lead has developed in them the rather annoying, and sometimes dangerous habit of showering passers by with sundry bits and pieces. This unfortunate habit once again underlines the need for a speed governor that is *reliable.* It may also suggest the desirability of siting the windmill away from areas where people often walk.

To conclude the chapter, Fig. 115 shows a rather delightful, and somewhat unorthodox, wind collector. We soon realize that a mere tree is a ready-made wind energy system; it stands up to the worst storms, can never be described as an eyesore and yet collects energy from an area of hundreds of square meters.

CHAPTER 16

Production and Use of Methane Gas

One of the world's greatest storers of Solar Energy is green vegetation. None of us know how it works, we have merely to accept the fact, and in our case concentrate on the recovery and use of the stored Energy.

In absorbing the Solar, the vegetation took the energy poor ingredients carbon dioxide and water, and built them into energy rich sugar compounds (carbohydrates) and oxygen. Virtually anything that has ever lived, animal or vegetable, contains much of this energy rich materials; our objective is to reverse the process to release the stored energy.

The more common of release processes, eating food, burning wood coal or oil, are quite apparent, the one we are interested in is the release of methane (marsh gas) from virtually any matter that has at one time been living and growing. (The term 'organic' covers once living matter, whereas, 'inorganic' is matter that never has lived, and includes carbon dioxide and water.) In the methane process this final return step is only taken when the gas is burnt, so leaving the original constituents carbon dioxide and water, plus the recovered Solar Energy.

The production of methane gas will appeal to fewer readers than will the collection of other Solar Energy forms. Methane production is something of a rural activity as evidenced by the lovely fertilizer it produces, but it is by no means a process limited to the 'out-of-towner'. Either way it is a fascinating simple process. We will look at two schemes, trial and family size, both of which follow the general form shown in Fig. 116.

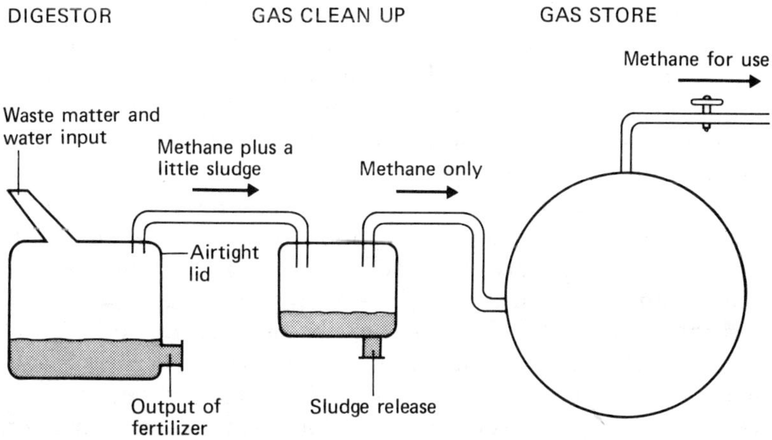

Fig. 116. Basic methane production and storage. The digester releases solar energy stored in vegetation by using friendly bacteria kept warm in an airless container to chew up the vegetable brew and release methane gas.

Methane gas is produced quite simply by feeding a slurry of waste vegetable matter, animal manure and water to be broken down in a digestor. Bacteria are introduced to start the process, and *warmth* and *airless* conditions maintained to continue it. Daily feedings of slurry are added with a not unpleasant residue (sludge) being removed periodically to prevent the process blocking up. Methane is piped off from the top of the digestor and passed through at least one stage of cleaning before going to store and eventual use. The sludge makes an excellent fertilizer.

Whilst there are very few suppliers of methane producing equipment in the world, plans are available via Directory contacts for what is mainly a self-construct involvement. But, even those who do not intend getting involved, it is interesting to see and understand how, with the aid of some friendly bacteria and the right conditions to encourage them, they are able to release Solar Energy that has been stored for months or years. Note how the energy 'release' is into a form that can be stored indefinitely, but is available for immediate use.

Whilst there are two basic types of methane production, batch and continuous, both of our schemes follow the popular requirement for continuous production. Fig. 117 shows our first scheme, which is really a working demonstration of continuous methane production. It does produce useable methane, but only sufficient to operate a single gas ring for 30 minutes a day. The scheme exactly demonstrates the principles that operate the larger schemes, and may provide the incentive for the building of the more ambitious scheme 2.

Both metaphorically and literally our scheme of Fig. 117 revolves around an old inner tube. A slurry (mixture) of water, manure and vegetable matter is fed in at the start of the circle, whereupon bacteria soon gets busy breaking it down into fertilizer and methane. The gas is allowed out at the top and the fertilizer is taken out at the end of the circle.

The construction of the digester starts by cutting completely through a truck or tractor inner tube at one point, and inserting a Plexiglas divider cylinder. The function of the cylinder is to provide a barrier inside the tube to prevent the slurry going round and round, and, to provide a mounting point for the three input and output tubes. The three tubes provide: input for the digestor, gas output and fertilizer output.

This is how the divider *cylinder* is made. Plexiglas sheet about 200 mm. (8 in.) wide is heated and formed into a circle to match the inside diameter of the inner tube. A disc of Plexiglas is cemented inside the circle as a center divider when it is finally formed and cemented into its cylindrical form. As the two cut ends of the inner tube will eventually be slid over the open ends of the divider cylinder (and are liable to slip off), a ridge is then added to each end of the cylinder to help retain them. Our cylinder complete with center divider and end ridges now has to have added to it three 50 mm. tubes. Bore three holes in the cylinder and insert the tubes as per the bottom of Fig. 117. The inlet tube is angled to encourage movement of matter round the tube; the short tube is flush on the inside to allow the methane out, and the outlet tube for the

Detail of three input and output tubes is omitted from main part of sketch to aid clarity

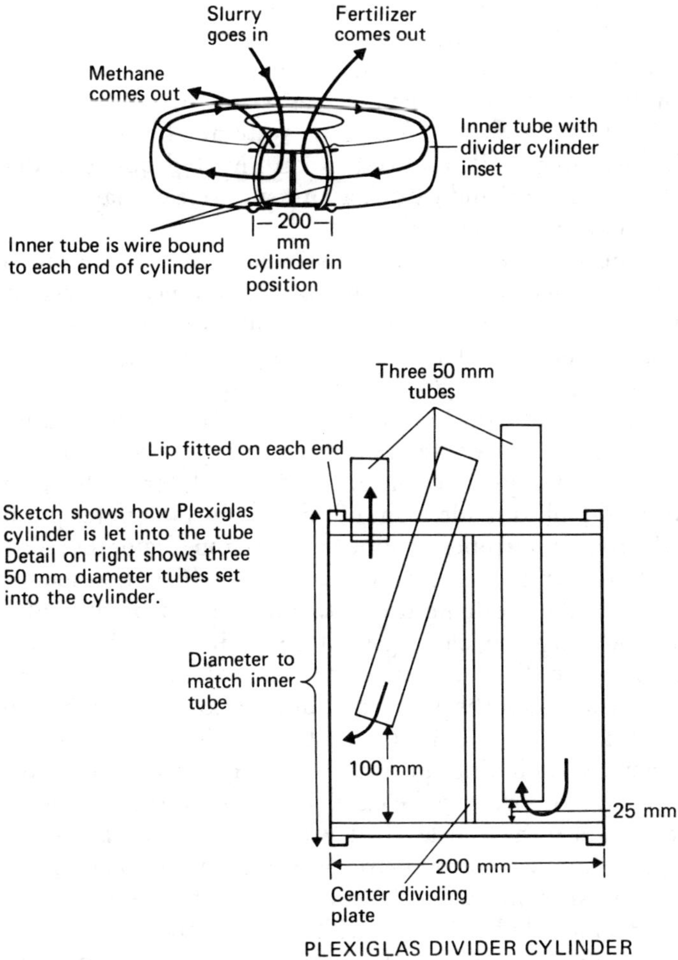

Fig. 117. An inner tube 'digester to produce methane gas.

fertilizer is kept close to the bottom to select the correct effluent.

The complete cylinder now can be added to the inner tube by slipping the cut ends over the ridges formed on the cylinder, and binding them in place with wire. In fitting the cylinder, the circle of the inner tube is complete once more (although blocked at one point); this also means that the digester is complete ready to take its place as part of the system.

Fig. 118 shows the complete system. The digestor now sits in a 'warming bed' followed by an inverted bottle 'scum trap', a water type 'gas indicator', a 'pressure release' jar, and finally a triple inner tube 'gasholder'.

In the complete system some means have to be provided to keep the digester warm, our sketch merely shows this as a warming plate surrounded with insulation, both of which are contained in a wooden box. Several possibilities exist to provide the heat, including Solar, but as a steady temperature is required it would be better to use gas or electric heating. One can consider lower temperatures of operation, but they do lower the gas production rate (details later). The pipe from the 'feed pail' is made from 50 mm. (2-in.) diameter piping with all the remaining pipes of about 6 to 10 mm. diameter (¼ to in.) The 'scum trap' is a simple device in which the scum carried over with the methane drops to the bottom of the bottle, thus allowing only the clean gas to go on its way. Next comes the sealed 'gas indicator' bottle, half full of water, another simple device that uses the sight of bubbles to indicate gas flow. An optional use of the liquid in the bottle is to 'scrub' the gas cleaner by including an additive in the water (see later). The open-topped 'pressure release' bottle that the gas next passes over, has a measured level of water above the open-ended tube to provide a certain pressure of water. In normal operation, the gas flows straight *across* the top of the bottle and does not enter the open-minded tube or the bottle. In the event of excessive pressure building up in the gasholder, the gas flows down the tube out of the open end, through the water, and out to the open air. The depth that the open-ended pipe is beneath the

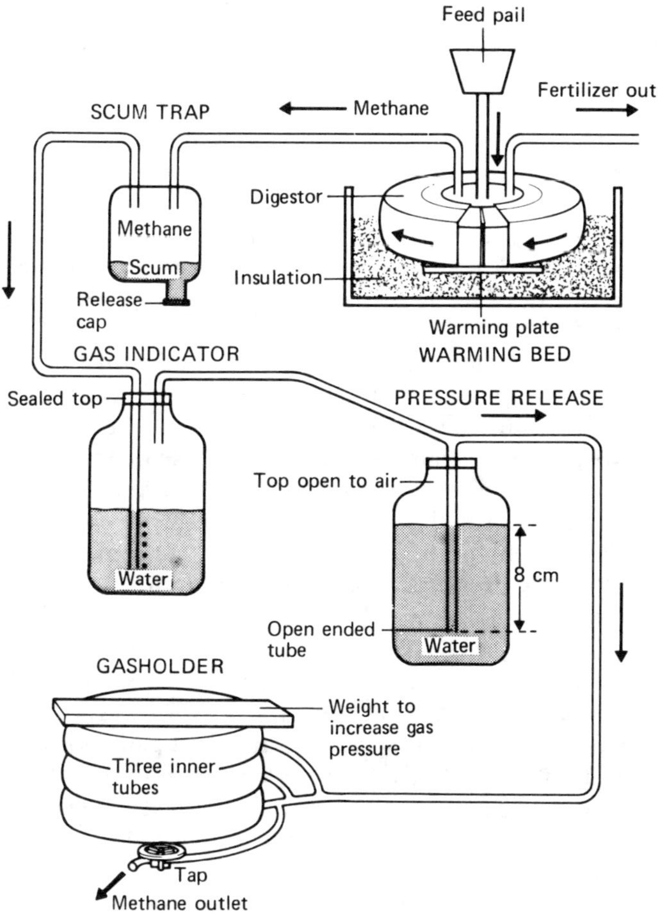

Fig. 118. A complete methane producing and storage system.

water determines the pressure at which the pressure relief starts. The 'gasholder' itself is made of three large inner tubes stacked one on the other with a weight placed on the top tube. Gas that has been produced and cleaned up is fed into all of the three normal inlet points of the tubes, the valves at the inlets having been removed. As gas flows in, pressure gradually inflates the tubes and eventually lifts the weight. The actual pressure arriving at the cooker, etc., is set by the amount of weight on top of the tubes. When the pressure weight has been set, the pressure relief level can be set by adjusting the depth of the open-ended tube. If gas is then produced in excess of the store capability, it will not attempt to overfill the holder but escape to the outside. A feed pipe and on/off valve completes the system.

By including lime water in the gas indicator instead of plain water, any small amount of the rather obnoxious hydrogen sulphide contained with the methane will be removed. If PVC or polythene containers or tubing are used instead of glass, they should be of the thick-walled variety otherwise methane will pass straight through them.

With the system built and its general functions understood we now go on to discuss the day-to-day operation.

To get the process going, the initial feed contains 5 kilograms (kgm.) (11 lbs) of active bacteria from the local sewage works, or a stagnant pond, added to the first daily feed of 1 kilo (2¼ lb) of manure with three liters (6 pints) of warm water. All this is loaded in on the first day via the feed pail and the digester warmed to 35°C. Daily feedings continue with the manure and water only. Be tidy with your daily feed and the overall smell of the system will be minimal as the decomposition starts up in the inner tube.

Decomposition is a term covering two processes. A normal compost heap, or open septic tank, operates with air present and is termed 'aerobic'; it produces heat as it works plus a rather unpleasant smell. We shall be using the other process, the 'anaerobic' or airless one. It has to have heat supplied to it, gives off no odor (it's sealed up anyway) and produces a fertilizer with a rather pleasant 'tarry' smell.

As our daily feeding of the digester continues, the process will gradually change itself from aerobic to anerobic as the air in the system is used up. As you can imagine, the start-up period is not typical of the normal operation as all the air in the pipes, bottles and tubes has to be used up; be patient with your system as it starves itself of air and goes anerobic. The first gas that comes through to fill the holder is a methane air mixture and should *not* be used, squeeze it out of the gas-holder to waste. Note that a methane air mixture if compressed to over 6.7 sq. cm. (95 lbs/sq. in.) will explode, as will a large volume of 5 to 14% methane in air mixture. Pure methane is a colorless, odorless, heavier than air gas that is normally quite safe. Remember that if there is a leak, any gas build-up starts at floor level as methane is heavier than air.

When the system has settled down to its correct anerobic state there will be approximately 70% methane, 29% carbon dioxide and possibly 1% hydrogen sulphide. If you leave the 1% in, by just using plain water in the gas indicator, the odor will aid the discovery of leaks. If you do not savor the smell of rotten eggs, then remove the hydrogen sulphide and introduce a small amount of tetrahydrothiosphere (your chemist will advise you) to give your methane that 'gas' smell. If you suffer the effects of water coming through the methane, then pass it through an anhydrous calcium chloride bed. The crystal bed will take in the water content, and it can then be regenerated (dried out) by heating.

By adjusting the weight on the top of the gasholder the normal pressure of 5.2 cm. (2 in.) water gauge can be obtained. The open end of the tube in the pressure release jar should be set at 5.2 cm. and the weight adjusted to just produce bubbles. Re-set the open-ended tube to 8 cm. on completion.

If you have a particular interest in the fertilizer effluent, it should normally be slightly alkaline with a ph of between 6.4 and 7.2, if it does get too acid this will show as foaming on the surface of the slurry (difficult to see in this first design), and a small amount of lime should be added to the slurry to correct the condition. If kept in good balance, the effluent makes an excellent liquid fertilizer, alternatively, it can be left to settle and dry in shallow containers. It produces.a crumbly, black, tarry smelling humus full of nutrients, with a well-balanced nitrogen, phosphate, potassium content ready for immediate use on the land.

If human excreta is used as part of the feed then a minimum period of 40 days in the digester, at a minimum temperature of 35°, has to be observed to ensure that any disease virus and pathogenic organisms are destroyed. As a significant last point; the fertilizer produced by a compost heap losses 50 to 75% of its nutrient due to oxidation within the heap, the increased value of the fertilizer produced by our airless method, with its *nil loss* of nutrients, should not be under-estimated.

After the process has settled down; the digester is happily full; methane is coming out the top and sweet smelling compost out the bottom; a gradual change can be made to the digester's

Table of Volatile Solids Produced Daily

	Wet raw		Volatile solids	
	kgm	lb	kgm	lb
Dairy cow	60	132	6.3	13.8
Dairy heifer	39	85	3.4	7.5
Beef feeder	27	60	2.7	5.9
Horse large	20	45	3.4	7.5
Horse med.	16	36	2.5	5.5
Pony	7	15	1.1	2.4
Hog breeder	11	25	0.72	1.6
Hog feeder	6	13	0.45	1.0
Sheep feeder	1.8	4	0.36	0.8
Lamb	0.7	1.5	0.09	0.2
Goose or turkey	0.27	0.6	0.045	0.1
Duck	0.18	0.4	0.03	0.07
Broiler chicken	0.14	0.3	0.023	0.05
Laying hen	0.09	0.2	0.018	0.04
Human, 68 kgm (150 lb)	1.2	2.7	0.11	0.23

Fig. 119. From each 1 kgm. of volatile solids, 370 liters of methane can be produced. An equal amount of wet weight (natural) vegetable waste can be added, for that assume 10% volatile solids only.

menu. Modify the diet to include 50/50 manure and vegetable matter. Chop up the vegetation into lengths of no more than 25 mm. (about an inch) before mixing it into the slurry.

Very large-scale digesters are surprisingly efficient and produce large amounts of gas. A good unit will produce a weight of gas that is one-third more than the weight of the so-called 'volatile solids' that it consumes. Whilst our units may not be up to this, they do produce gas quantities related to the amount of volatile solids consumed.

From the 1 kgm. of laying hen manure used daily in the inner tube digester, 750 grams is water, leaving 250 grams of solids, of which 200 are volatile solids (V.S.) and are the only part that can be used to produce methane. Fig. 121 shows a good range of possible sources of V.S. as daily output per animal. Given that 1 kgm. of V.S. produces about 370 liters of methane (1 lb of V.S. produces 6 cu. ft. or, 1 lb. wet weight of manure produces 1 cu. ft.) estimates of gas production are readily made. Our little tube digester will produce 74 liters of methane a *day* (about 2½ cu. ft.).

From the next table, Fig. 120, the amount of methane required to achieve various functions is also shown. The diesel engine converted to methane operation is of particular interest. In common with many engines it has a quite low efficiency

	Liters	Cubic ft.	
Lighting	70	2.5	per mantle per hour.
Cooking	113/226	4/8	per hour per 4-inch ring.
”	340/420	12/15	per person per day.
Gas engine (25% effic.)	450	16	per brake horsepower per hour.
Gas fridge	34	1.2	per cubic ft. of space.
Incubator	17	0.6	per cubic ft. of space.

Fig. 120. How far the methane goes. If waste heat is fed back to keep the digester warm the whole operation is much more productive.

(actually 25% is quite good) and must, therefore, put out 75% of the energy as heat. By siting the engine next to the digester most of the heat can be re-cycled back to keep the process warm.

The actual heat value of the gas is 0.007 kw/hr per liter, that is 7 watts per hour from each liter (680 BtU/cu. ft.). This makes our *daily* output from the little digester a little over 0.5 kw/hr (500 watts for an hour).

At last we are back to recognizable terms, and a restoration of faith that the plants did know what they were doing all the time!

If some of the gas is used to keep the digester warm it will use 35% of that produced if the unit is working in a temperate climate. In a cold climate 100% of the gas will be used but one is still left with that lovely fertilizer.

Thus far we have dealt with the construction of an experimental size of methane producer; established the amount of dung required; and seen the heat value of the gas. We now look at a practical system to suit a 4-person household. Unlike Solar ray collection, methane production does benefit from the normal rules for scaled-up systems. Bigger is better this time.

The size of a system is often governed by the amount of V.S. available on a regular basis, and assuming that it does not produce more gas than is required, it is a perfectly reasonable pointer. The second system is based on a cylindrical tank lying on its side, we choose a 2000-liter capacity but this can be scaled up or down to suit your needs without upsetting the operation. Work out your daily feed size on one part wet weight of solids to nine parts water, and allow space for 50 days' loading to give you a full digester, or half full if you opt to use the top half of the digester as storage. The 2000 liter (70 cu. ft. or 500 U.S. gals.) is sufficient for a 4-person household including the possibility of some gas storage in the tank.

Let us visualize a situation where the daily feed is three times the previous rate and that the stage had been reached where the feed was 50/50 vegetable matter, 1.5 kgm of chicken manure, and 27 liters of water. A 50-day cycle means that 50 times the 30 liters has to be stored in the tank, or three-quarters full. From the solid matter we get:

Manure solids	1.5 kgm. = 0.3 kgm. V.S.	
Veg. solids	1.5 kgm. = 0.15	
Total V.S.	0.45 kgm.	

this will produce 166 liters of methane per day, giving an equivalent heat energy of 1.2 kw/hr. This would increase to 1.5 kw/hr if all manure was being used.

Having established a reasonably sized device for a family, we should now look at the system a little more closely. The tank should be left as shown, lying down, as this gives a larger surface area for the slurry inside, the reason is that scum, which always forms, will then be distributed over a larger area and have less of a clogging effect. For the same reason, the surface area should not be reduced too much, by filling the tank more than three-quarters full. As the system is required to be kept close to 35°C, a means of heat, and a layer of insulation are also required. Much of the heat could come from Solar, waste domestic water, wind electric, waste heat from an engine, or even by putting a compost heap around the digester. In most cases some form of back-up has to be provided from an electric or gas heater, we have shown the heater inside the tank, it could also be outside.

If one does allow the temperature of the digester to drop and remain at 30°C, the process takes 2 months to digest the input materials instead of the usual one, and the daily gas production rate drops by one-half. If the temperature is held at 25°C it is then a 3-month process with a one-third gas production rate. In neither case is anything lost but the time scale is extended. If different temperatures are considered to suit your own situation, the important thing to avoid is a sudden temperature change as this appears to upset the bacteria.

The cheapest way to insulate the digester is to sink it into the ground after first lining the pit with a generous layer of straw or sawdust. Do not cover the top of the tank with earth as you are bound to require access for umpteen reasons before long. Cover the top with a thick but moveable layer of insulation.

One other factor to ponder, with this family size system, is whether to include human waste products in the digester! In addition to the earlier requirement about a longer digestion period, the decision often revolves around the degree of complication involved in altering your existing plumbing, plus your personal view of the matter. As can be worked out from the table, a family of four produce 0.4 kgm. of V.S. per day to produce about 150 liters of methane.

Taking all the factors into account that we have discussed, a practical configuration for the digester is as Fig. 121. If you have the benefit of having made the previous model you may well come up with some improvements and variations, on the other hand, if you require more detailed plans of other and/or

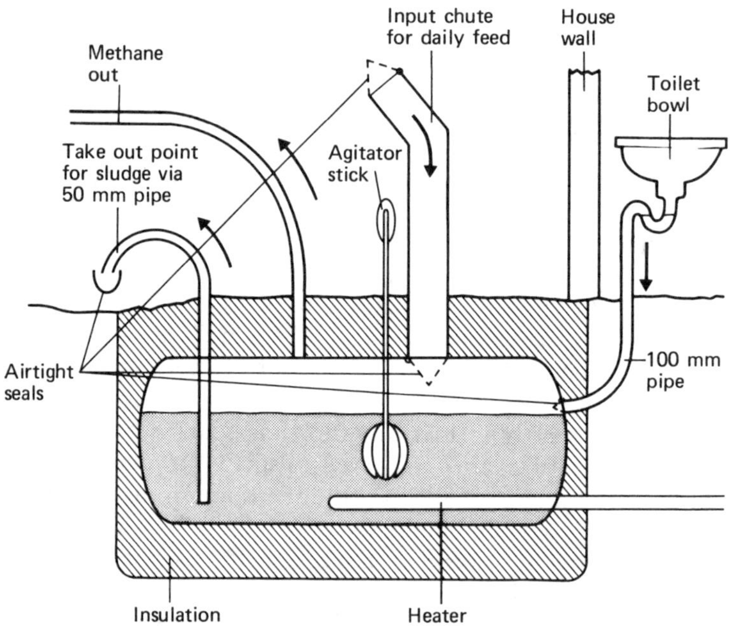

Fig. 121. A four family size digestor buried below ground to retain heat. The toilet feed is optional.

larger systems, these are available via the Directory.

In our family size digester the agitator device shown should be used straight after the daily feed of new material has been added. If this is all manure then 5-minutes' agitation is sufficient, but more will be required if vegetable matter is included. Do not install a *continuous* form of agitation as this appears to upset the bacteria – you will have gathered by now that they are somewhat temperamental little fellows.

It is a good but subtle improvement to tilt the tank slightly (5°), making the end away from the input the low point. This small tilt helps the general progression of matter from end to end. New material goes in at the high end and sludge is taken out from the low end. The digestion process, as it proceeds, gradually stratifies the slurry into layers as shown in Fig. 122. The inorganic solids can either be taken out with the fertilizer or emptied periodically and dumped. Returning

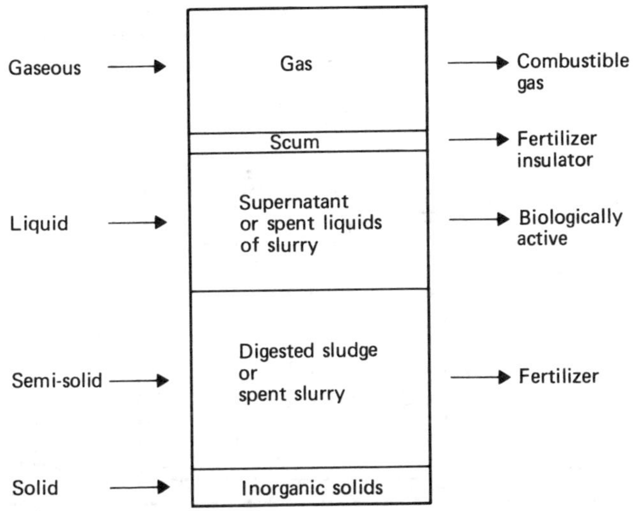

Fig. 122. The digestion process slows down as the slurry moves through and stratifies out to the layers shown here.

to Fig. 121, it can be seen that all air is kept from entering the process by fitting single or double seals over inlet and outlet points. The heater is at its most efficient as shown but can operate on the outside. Methane is taken out of the top area of the tank (which can also be used as a store) and used direct from there, but more normally it is fed via a series of stages similar to the first system we discussed. With the amount of gas that we have estimated will come from this larger unit, most of the previous equipment will suffice, but there are two areas that can be improved.

Fig. 123 shows an improved gasholder arrangement. As gas bubbles up through the water on the right side and into the metal drum, pressure builds up in both drums to gradually lift them out of the water. The space above the water provides storage for the gas, and the weight of the drums provides pressure to make the gas managable. If the drums are too heavy a counterweight will have to be fitted otherwise the high back pressure felt by the digester will reduce gas production. Aim to work at the same pressure as before (measured on the water guage as 5.2 cm) as this is the norm. The capacity of the drums

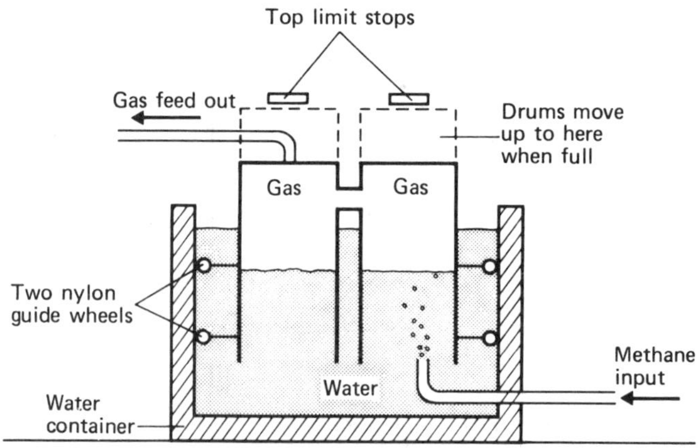

Fig. 123. Two metal drums floating upside down in a tank makes a large holder.

should total 1000 liters (35 cu. ft.) to hold 6 full days of gas production from a family size system. The top limit stops prevent the drums from coming right out of the water. If gas production continues after the drums have hit the stops, pressure builds up until it escapes to the air by way of the pressure release jar.

Fig. 124 shows one other device to make the operation of the family sizc digester easier. A simple lever operated pump with a series of valves can be made to carry out several tasks. One, the daily input slurry can be pumped in and not poured. Two, sludge can be pumped out and not carried. Three, the digestor contents can be circulated to effect the necessary agitation.

A poetic story goes the rounds of the methane fraternity concerning a rather stalwart enthusiast who put *all* his waste products into the methane system. One winter after a somewhat protracted bout of flu, our hero was induced to take a heavy dose of antibiotics, and it was not until some weeks later, when his methane production abruptly ceased, that he realized just where the antibiotics had passed to – and what they had done to his friendly bacteria!

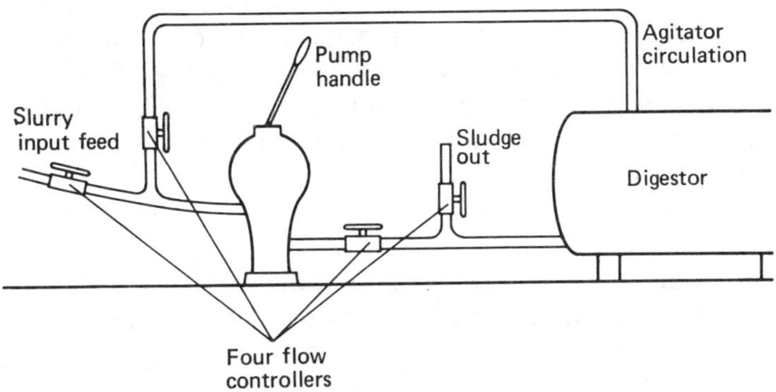

Fig. 124. One lever pump that by altering the flow controllers can be made to do three different jobs.

CHAPTER 17

The Economics of Solar Energy

Nobody with a broad knowledge of Solar equipment will now deny that Solar is economically viable; equally so few will contend that all systems are viable. That some systems are viable and others are not makes it doubly important to know the true worth of any system.

Our examination of costs considers the basic factors that apply to all installations: the cost of Solar equipment, a fair return from the investment, the general rate of inflation and the cost of maintenance. The methods used are equally applicable to passive or active systems, new homes or retrofit systems, wind or methane systems, do it youself or buy in, and the different Sun areas.

In trying to answer the question, "certainly Solar is great but how much does it cost?", this poses some even trickier questions. How much will oil cost in 10 years' time? etc., but having no intention of ducking out, we have inserted today's best estimate and proceeded to answer the question. The substitution of a more pertinent figure at a later date allows a good up-dated figure to be obtained.

For those who insist on a YES/NO answer to the cost question, Fig. 125 provides just that.

Our more detailed cost examination, yet still in lay terms, shows three examples, quite sufficient to establish a pattern that answers the question, 'should I invest in Solar?'.

At the outset the following concept should be understood. If a person bought a car that never required fuel, he would supposedly expect to pay more for it in the first place. Likewise with Solar; the initial outlay is greater but the running costs are

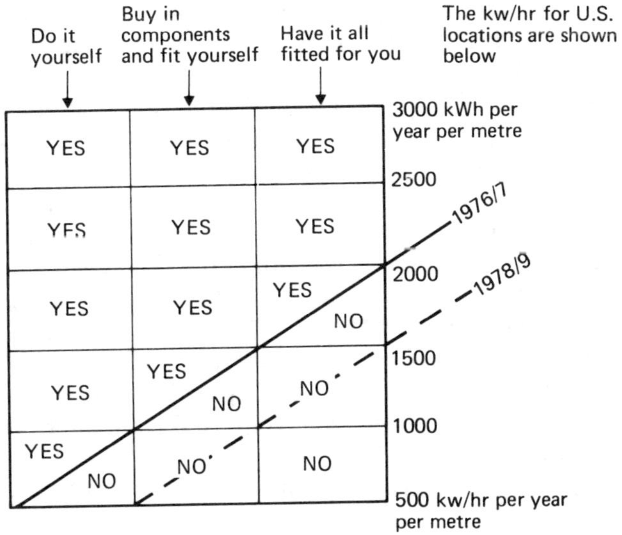

Fig. 125. Is solar cheaper? The diagonal line shows the changeover dates in the 'maybe' blocks. Three had a 1976/7 charge with the next two in 1978/9.

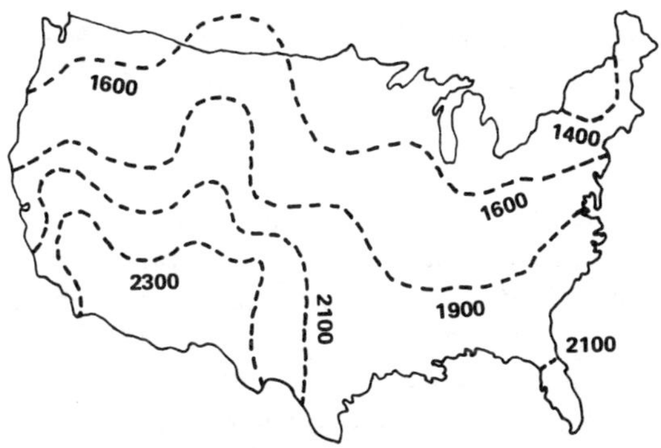

Fig. 126. Approximate total kw/hr. per year that arrive at various U.S. locations. The end of chapter 8 gives more detailed data.

negligible. By comparing Solar and non-Solar costs at the outset one will arrive at the wrong conclusion. The following 4 steps will lead to the correct conclusion.

(1) A starting point is established by defining a home in terms of its Sun energy input, size and occupancy, present cost and method of heating water and the home.

(2) A suitable Solar system is envisaged and costed together with the likely Solar contribution it could make.

(3) The operating costs are projected for a Solar and non-Solar system over the next 20 years. (Twenty is the expected life of a system.)

(4) Total up the difference in cost for each year of operation to establish the 'breakeven' year, and, the total saving that could be expected over the life of the Solar equipment.

The projected saving over the years is represented by the formula:

Year 1 costs= $(\text{fuel})+(\text{electric})+\left(\dfrac{\text{net cost of system}}{20}\right)+(\text{maintenance})$;

Year 2 costs= $(\text{year 2 fuel+MFI})+(\text{year 2 electric+MFI})+\left(\dfrac{\text{net cost of system}}{20}\right)+(\text{year 2 maintenance+inflation})$;

Year 3 costs= $(\text{year 1 fuel+MFI})+(\text{year 1 electric+MFI})+\left(\dfrac{\text{net cost of system}}{20}\right)+(\text{maintenance+inflation})$;

etc. on to year 20.

The above formula is used to produce sets of figures for the conventional and Solar systems. In example B, at end of this chapter, you will see that the 4 items in year 1 costs are worked out as

$285 + 208 + 17 + 20 for the conventional,

and $185 + 40 + 65 + 30 for the Solar.

What will be seen is that whilst Solar can be more expensive in the first years (in the above example it is not), if the invest-

*See point 5 following.

ment is added in, the constantly recurring rise in fuel costs of the conventional syste, make it more expensive after perhaps 6 or 8 years. After this time the Solar system is paid for and continues to run for next to nothing, whereas the conventional system is costing more and more to run every year *ad inifinitum.*

Another very valid argument in favor of investing in Solar is that one does have an asset to show for the outlay. If one buys $100 worth of fuel oil, and burns it, it is gone, whereas, $100 worth of Solar equipment is still there even if one wishes to sell the house!

In our three examples, the selected format allows an easy re-work to your precise situation. As the larger methane system was taken fairly well through, a further economic appraisal is not included here.

There are several points to define before we use the above formula in the three examples.

1. A fair return for the investment in Solar equipment is perhaps the crux of the whole exercise, therefore, it is that we tackle first. The simplest approach is merely to allow a 'straight depreciation' of the investment over the life of the equipment. If $2000 has been invested and the life is 20 years, then the $100 a year that drops off of the value should be viewed as the amount of fuel that has to be saved in the year to merely break even.
2. The second method is more realistic as it allows for what economists love to call the 'time value of money', a delicious phrase that, as far as we are concerned, means that we should allow for the interest that the cash would earn if it had been left in the bank.
3. It is a rather convenient characteristic of some rather tiresome monetary calculations (and you will have to take our word for this), that if an interest rate of 10% is applicable, and a 10% inflation rate is allowed for, then the two cancel out. And at the same time we can return to the straight depreciation method we started with and still be realistic.
4. Whilst interest rates and inflation rates tend to go hand in

hand, fuel inflation rates are certain to be higher. This is so for at least three reasons. One: as any item runs out as fossil fuels are, they always command a scarcity price. Two: the easily, and therefore cheapest, obtained portions of the fossil deposits were used first. Three: the multiple problems of the 'cheap' replacement, atomic energy, are inflating its costs faster than any other fuel. We ignore any short-term political tinkering with these basic premises.

5. By accepting the certainty that fuel cost inflation will be greater than general inflation, we can take the difference between the fuel cost inflation rate and the interest rate (both as percentages) and use it as our 'modified fuel inflation rate' (MFI rate). In the case of an interest rate of 8%, and a fuel inflation rate of 20% we can use the difference of 12% as our MFI rate, and 'straight depreciation' for the equipment. We use this third method in our three examples.
6. In selecting a Solar system one should not normally attempt a 10% or 100% Solar contribution; the percentage that gives the best return will generally lie in the 50% to 75% region.
7. Whatever the expenditure on a system, U.S. users can deduct 25% of the cost of Solar equipment as a tax credit (providing one pays enough), this brings \$4000 worth of equipment down to \$3000.
8. Additionally, one should allow for the equipment, both Solar and conventional, to have some value after 20 years. This will either be as scrap, or, because part of the system can be re-used. Therefore, it is more realistic to depreciate down to 25% of the installed price. This is not an over-generous figure as the value of what remains will be judged against the then year 2000 prices!
9. After deducting the two previous allowances, the balance should be divided by 20 to give the 'yearly depreciation'.
10. A 'yearly depreciation' figure has also to be worked out for the existing system, and as a wet system (steam or

water) will last 25/30 years, and a dry one (hot air) 15/20 years, a good average for a conventional system would still be about 20 years. Allowing for the end value of the equipment, the remainder should be divided by 20 to give the depreciation.

11. In the examples we use a yearly fuel inflation rate of 20%, so that after subtracting a 10% interest rate, we have an MFI rate of 10%. (A 10% tax-free interest rate is quite reasonable for a fixed period investment of 20 years.)
12. Up-dated information on inflation rates are available from Federal statistics on display at any Public Library.
13. A first estimate of 'energy produced per year' is given by the following. Multiply the local kw/hrs (for the area in question), arriving on 1 square meter, by the number of square meters of collector area, and reduce the product to 40% of itself to obtain the answer. (This 40% represents a 50% efficient collector and further losses in the remainder of the system.)
14. A more accurate estimate of the amount of energy produced per year is obtained by comparing the monthly fuel used V the monthly energy collected. A further refinement can be achieved by differentiating between fuel for heating (natural gas or oil), and electricity for water heating in the summer. An example of a yearly energy budget is shown in Fig. 127. In the construction of your own graph you have first to ascertain the monthly use of each type of fuel. With monthly billing this is easy, if not, it will require either monthly monitoring or intelligent apportionment to provide the detail. The expected Solar input can be obtained from the data given in Chapter 8.
15. It is beyond the scope of this book to detail how to *measure* the Solar input (although a direct method such as this is always preferable to getting a figure from a graph), but the basis of the measurement can be given. The basic requirement is to measure the *difference* in temperature of the heat as it arrives and departs, then, to

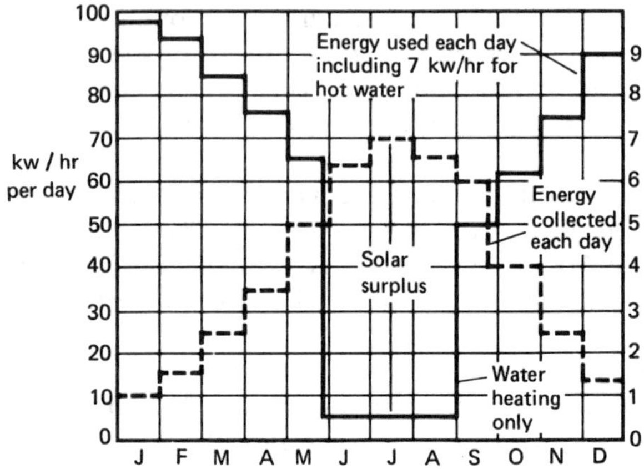

Collected 11 15 25 35 49 66 71 68 57 42 27 14 = 480

480 by 30 days = 14,400 kw/hr in the year (8730 useable)

Used 98 95 85 76 67 7 7 7 52 62 74 90 = 720

720 by 30 days = 21,600 kWh in the year,

(19,000 for heat and 2600 for water heating)

Fig. 127. The energy budget of 'Example B' to follow. Note how the peak of summer solar input exceeds the needs of June, July, August and September.

multiply this by the *volume* of the delivery medium that passed whilst making the temperature measurements. For example, if a cubic meter of water entered the heating system at 40°C (104°F), but left at 30°C (86°F); the 1 cubic meter of water lost 10°C and consequently delivered 8.5 kw/hrs. The reverse calculation is also true. If a collector raises a 2-cubic meter tank by 5°C it has delivered 8.5 kw/hrs to the tank. (There can be complications here as the tank may be giving out heat to the home at the same time.)

16. The cost of equipment in the examples is equally applicable for new home installations, or, for retrofit to

existing homes. It is an interesting exercise to see whether the additional loan charges (if one raised the cash that way) incurred in purchasing a Solar system are paid for by the reduction in fuel charges. As at least the interest charges are deductible (and fuel bills are certainly not), the results could be surprising. Additionally, the asset will be clear of capital gains tax.

17. Fuel oil number 2 has been costed at 55c a U.S. gallon, including taxes and delivery charges. For reference: this oil has a heat value of 40 kw/hr (138,000 BtU.) per U.S. gal., and when burnt at a normal boiler efficiency of 70% will produce 28 kw/hr of heat for the home. That's virtually 2c per kw/hr.
18. Natural gas is costed at 30 c per therm (100,000 BtU.) including taxes, and will have a heat value of 29 kw/hr. When burnt at 70% efficiency this will deliver 20 kw/hr to the house at 1½ c per kw/hr.
19. Electricity is costed at 8 c per kw/hr including the very considerable 'fuel adjustment' charge, and taxes. Thankfully, electricity is 100% efficient at the point of use.
20. Maintenance is a recurring charge that will inflate each year at the general inflation rate. One could also expect that if Solar took over half of the work, the conventional boiler would require less maintenance, and thus incur less charges. You may even choose to ignore the maintenance item altogether by assuming that the combined charges will be the same as they were before the Solar was added, each now doing half of the work!
21. The Sun's energy is not expected to rise in price.

Now for the examples. As you will soon appreciate, it is the regional cost of fuel as much as the amount of Solar in the region that decides the viability! Please remember that our objective is to set a pattern to enable you to substitute your situation, rather than create controversy over the actual figures we have used.

Example A. A detached house of 150 sq. meters (1600 sq. ft.) with 4 in family, oil-fired heating, situated in 2000 kw/ per year Sun area.

Cost of Solar equipment (home heating only)

Collector of 30 sq. meters at $80 per meter	$2400
Ancillary equipment	200
Storage tank	300
Piping	100
Labour – nil, as it was self-installed.	
Total cost	3000
Allow 25% tax credit; expenditure reduces to	2250
Subtract 25% for the residual value; balance now	1500

Divide balance by 20 to give deprec/yr. of $75

Cost of existing equipment

Estimated total cost	$640
Subtract 25% as the residual value; balance now	480

Divide balance by 20 to give deprec/yr. of $24

Energy produced per year

Simplified calculation. 2000 by 30 by 40% = 24,000 kw/hr/ yr, but as can be seen from Fig. 127, you would be wise to reduce this amount by 40% as much of the summer energy cannot be used.

Useable energy collected 14,400 kw/hr.

Energy used per year (oil)

Normal yearly use of 3800 liters (1000 gal.) produces 28,000 kw/hr. and costs $550.
With a 50% Solar contribution this will reduce to $275.

The electricity for water heating (domestic) will remain the same, as the system was not designed to contribute.

The saving over the 20 years can now be set out as previously given in the formula on page 200.

Costs in year 1

	No Solar	*With Solar*
Oil	$550	$275
Electric	no change.	
Deprec/yr	24	75
Maintain.	30	40
Total =	$604V	$390

Saving in year 1 = $214

Costs in year 2

Oil	$605	$302
Deprec/yr	24	75
Maintain.	33	44
Total =	$662V	$421

Saving in year 2 = $241

Costs in year 3

Oil	$665	$332
Deprec/yr	24	75
Maintain.	36	48
Total =	$725V	$455

Saving in *year 3 = $270*
year 4 = $302
year 5 = $336
year 6 = $376
year 7 = $418

At this time the accumulated saving is $2157 or virtually all of the $2250 invested. This is a 7-year payback period.

And the savings continue:

year 8	$464
year 9	$516
year 10	$575
year 11	$637
year 12	$705
year 13	$781
year 14	$866
year 15	$956
year 16	$1055
year 17	$1167
year 18	$1289
year 19	$1422
year 20	$1568

At this point the accumulated saving amounts to just over $14,000 in 20 years.

Example B. An apartment of 3 rooms, 2 in family, individual heating for each room by natural gas, water heating by electric. Sun strength of 1800 kw/hr/yr.

Cost of Solar equipment to heat apt. and water

Collector of 20 sq. meters at $90/meter	$1800
Ancillary equipment	200
Storage equipment	200
Piping	100
Labour	300
Total cost	$2600
Allow tax credit of 25%; expenditure reduces to	1950
Subtract 25% for the residual value; balance now	1300

Divide balance by 20 to give deprec/yr of $65

Cost of existing equipment

Estimated total cost	$440
Subtract 25% for the residual value; balance now	330

Divide balance by 20 to give deprec/yr of $17 (approx)

Energy produced per year (previous Fig. 127 applies).

As the graph of Fig. 127 shows there is a summer surplus of Solar of 5670 kw/hr (189 times 30 days); this leaves useable energy collected of 8730 kw/hrs.

Allowing for 100 liters of hot water per day (as 1 kw/hr will raise this amount, 8.5°C, 7 kw/hr a day will be required), in a year 2600 kw/hr are needed. But as the collector will not reach 60°C in the winter not all will come from Solar.

With no Solar, Costs are:	2600 kw/hr of electric for water at 8c	= $208
	19000 kw/hr of gas for home heat at 1.5c	285
	21600 kw/hr *total energy* costs	= $493
With Solar, costs will be close to:	500 kw/hr of electric for water at 8c	= $40
	2100 kw/hr of Solar for water	nil
	6630 kw/hr of Solar for home heat	nil
	12370 kw/hr of gas for home heat at 1.5c	$185
	21600 kw/hr *total energy* costs	= $225

The costs over the 20 years now appear as:

Costs in year 1

	No Solar	*With Solar*
Gas	$285	$185
Electric	208	40
Deprec/yr	17	65
Maintain.	20	30
Total =	$530V	$320

Saving in year 1 = $210

Costs in year 2

Gas	$313	$203
Electric	229	44
Deprec/yr	17	65
Maintain.	22	33
Total =	$581V	$345

Saving in year 2 = $236

Saving in year 3 = $265
year 4 = $302
year 5 = $329
year 6 = $368
year 7 = $408

At this time the accumulated saving is $2111 giving a small excess over the original investment. Another 7-year payback period.

And the savings continue:

year 8	$455
year 9	$506
year 10	$560
year 11	$621
year 12	$688
year 13	$761
year 14	$842
year 15	$932
year 16	$1029
year 17	$1138
year 18	$1256
year 19	$1401
year 20	$1531

At this point the accumulated saving amounts to just over $13,800 in 20 years.

Example C. A country home of 200 sq. meters (2150 sq. ft.), 6 in family, domestic water heated by electricity, average wind speed 14 m.p.h. (coastal area).

Cost of windmill (water heating only)

Windmill and generator of 1 kw. rating	$700
Stub tower (self-built) on outbuilding. Materials	$ 50
Immersion heater and wiring	$ 50
Total cost	$800

Allow 25% tax credit; expenditure reduces to	$600
Allow 10% only as the residual value, balance now	$540

Divide balance by 20 to give deprec/yr of $27

Energy produced per year

From the second table in Chapter 15 we can see that a 1 kw. generator will produce 104 kw/hr per month from a 14 m.p.h. average wind. With 6 in family, 21 kw/hr is required for water heating, therefore, all the windmill energy can be used.
Using all the energy offsets the electricity by,

104 by 12 months by 8c = $100 in year 1.

With the wind system only altering the cost of electricity in the house, the savings are:

year 1 $100 less $27 for deprec/yr	= $73
year 2 $110 less $27	= $83
year 3 $121 less $27	= $94
	= $106
	= $119
	= $134

At year 6 the savings exceed the $600 expenditure. The total savings through to year 20 exceed $5000

CHAPTER 18

Future Developments of Solar Energy

A story from a not too reliable source tells of an OPEC gentleman who, as he was sitting one day munching on a hamburger and sipping a coke, suddenly realized that his check for $ 2.50 exactly matched what he was getting for a barrel of oil! Whether that incident actually triggered the oil crisis no one will ever know, but we do know for certain that cheap fuels are finished and that Solar is one of its future replacements.

Unfortunately, many of the future developments of Solar Energy are being aimed to use it in large installations. Giant Solar farms are proposed covering many acres, orbiting satellites beaming large amounts of energy to earth, Solar forests of fast growing trees feeding a power station in the middle, etc.

All these schemes are an unfortunate slavish following of the present power generating system that was itself based on *concentrated* energy sources. Whether it was coal, oil, gas or nuclear, the system dictated that they all had to be burnt at a central point to produce heat and steam to drive the electric generators. A vast distribution network then had to be created to spread the power out as best it could. With only a 40% efficiency at the power station, and a doubling of its cost for every 50 km. (30 miles) it was taken, an 8 c per kilowatt price – as compared with a 1½ to 2 c price if burnt locally – can easily be understood. When a *ready distributed* form of energy is available and a course of action suggests that it is all gathered up and then spread out again – that is difficult to understand.

We are ignoring all these schemes with a sense of bewilderment (unless they are proposed as an adjunct to home Solar), and instead concentrate on individual systems; systems that retain for the user that most cherished desire of being able to control both price and availability.

Before discussing some likely new developments, there is one important overview consideration that will become more and more important in the future in relation to any energy source. In the long term, full consideration will have to be given to the *amount* of energy that is required to produce more energy. It can be readily appreciated that, if it takes 100 kw/hrs of energy to mine 95 kw/hrs of coal energy, then it is better to leave it in the ground. Less evident, is the increasing problem of the amount of fertilizer type energy that is being required to produce food energy. As more and more fertilizer is now required to grow the same amount of food, are we putting that energy to the best use on the farm?

The two most active *new* areas for our type of Solar Equipment will be Solar cooling and heat pumps: two processes that are in effect one and the same. Solar cooling is such a natural use of that energy that the problems are bound to be solved before long. We then have the delightful situation where the supply and demand are in step. The hotter it gets outside, the better the Solar cooler works inside. Ideal.

Fig. 128 shows the popular compression system as is used in refrigerators, air conditioners and heat pumps.

As a cooler: heat is absorbed by the evaporator *inside* the house and released by the condenser on the *outside.*

As a heater: heat is absorbed by the evaporator *outside* the house and released by the condenser on the *inside.*

A second popular method, although basically less efficient, is the absorption system, again it is one that can be used for heating and cooling. In this method the three basic elements, condenser, expansion valve and evaporator are identical with the previous method, but the compressor is replaced by a more complex flow system. The energy then applied in place of electricity or mechanical rotation to drive the compressor, is a steady supply of heat. This now gives us two possible systems with three forms of energy to operate them.

Of the three possible ways of operating these heat pumps or cooling systems from Solar Energy, the least favored is by an electrically driven pump. This is so because of the several conversions of the energy before it does its work of compressing the refrigerant.

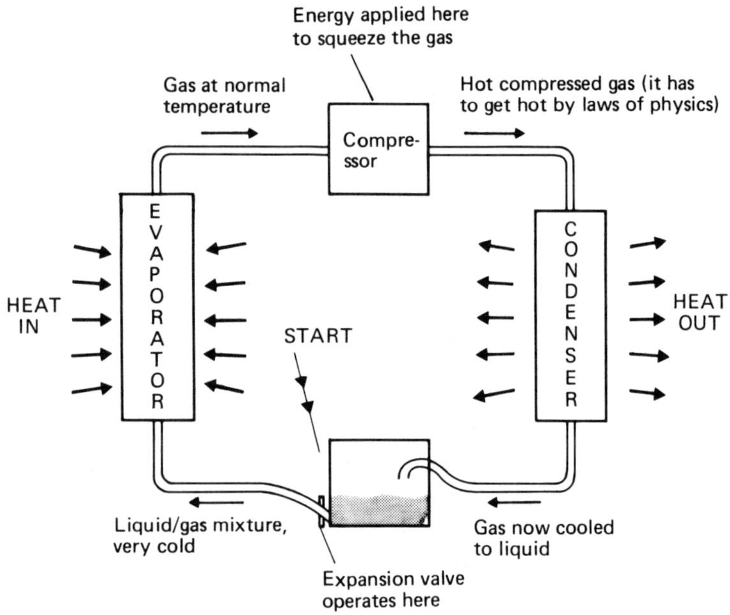

Fig. 128. Compression refrigerator or compression heat pump – its the same! The circulating refrigerent acts like a sponge. Let it expand, as on the left, and it soaks up heat; squeeze it, as it goes to the right, and it releases the heat. This way of *moving heat* can be used to cool a room by moving heat out; or heat a room by moving the heat in.

Of course, heat pumps can be driven from purchased electricity, but there is a problem. Unless a heat pump has a supply of bulk heat above about 15°C, its C.O.P. (efficiency) will drop below 4. And whilst a heat pump is as close as one will ever get to a 'something for nothing device' (with a C.O.P. of 4 it will actually give out 4 kw/hr for every 1 kw/hr of electricity fed in), as the cost of electric is 4 times the price of oil there is little progress. The likely sources of this bulk heat are: the outside air, which will be below 15°C when heating is required; a stream which will be at about 10°C; and heat from below ground at about 12°C. All heat pumps do this supposedly impossible trick by taking a small amount of heat (perhaps 1°C) from a large volume of heat and transposing it into a smaller volume of much greater heat.

The Solar fraternity are now concentrating their efforts on adapting the well-tried compression and absorption schemes to Solar operation. First efforts apply steady Solar heat to the absorption system in place of gas heaters. Valiant work by Professor Eric Farber and his team at the University of Florida already have units of 5 ton cooling capacity working well. Other work has integrated the collector into the system by flowing the refrigerant through the collector! A device with a 1½ sq. meter collector produces 36 kgm. (80 lb) of ice on a good day. These figures alone show the promising capabilities.

A second adaption to Solar operation concentrates on the compression system both as a heat pump and as a cooling device. This time simple mechanical rotation is applied from a windmill or waterwheel in place of the electrically driven compressor. There are problems due to the intermittent nature of the wind but these are being ironed out with flywheels and improved refrigerants. The direct use of the energy from a windmill in this way is a very efficient system of great promise.

Even greater benefits can be imagined by the obvious trick of using the same set of equipment to achieve winter heating and summer cooling. To obviate physically changing over the condenser and the evaporator, the units are being designed as dual purpose devices. Little more then remains than to reverse the refrigerant flow to change from a cooler to a heater. Sounds marvellous.

A simple idea was recently proposed to provide a supply of *higher* temperature bulk heat to feed a heat pump. All that is required is to remove a section of tiling from a house roof and replace it with some form of transparent material. The material could be matching glass tiles, or a flat sheet of glass/plastic – it will look like a skylight. The Solar Energy now let into the roof loft heats this bulk supply of air up to a point where it is worth switching on the heat pump. The pump gathers up the loft heat and transfers it to the Solar store. With a wind operated heat pump the saving would be terrific.

By elaborating on the original loft idea other sources of bulk heat can be found. Fig. 129 shows some possibilities. Collector 4 is a source of heat that may be found between the layers of a cavity wall as the day's heat soaks through. If this is filled with insulation, as it should be, or too much warmth is taken from the inside of the house, an extra layer

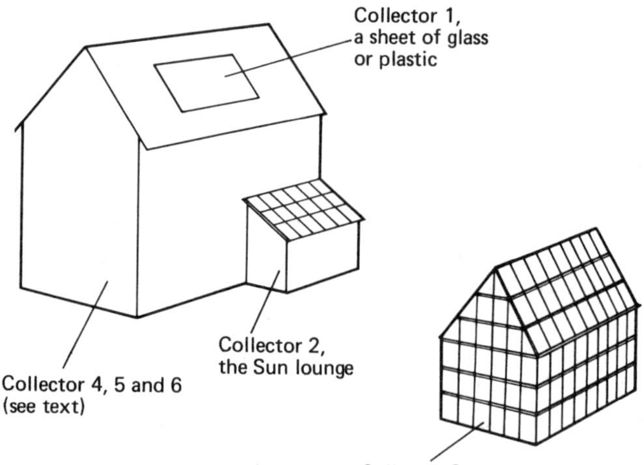

Fig. 129. If you have neither of these outbuildings then let a window into your roof and collect bulk heat in the loft, and then heat pump it into store. If you have a greenhouse or Sun lounge do not open the window when they get too hot inside, heat pump the surplus into your store. With those temperatures you get energy at 1 c per kw/hr even if you drive it from 8 c electricity.

could be formed by building a wall (with an air gap) in front of the existing wall. The heat can then be taken from between the two, which must be better than taking it from the outside air. But better still, form a spaced glass or plastic wall and take the heat from between the glass and the existing brick. As you see Solar ideas tend to proliferate rather freely – by now you may well have had some of your own! It is ideas like the 'no cost' loft collector that pulverize the arguments that Solar is too expensive.

Having seen the need for mechanical energy, other than from wind power, there is hope that alternatives will be forthcoming to provide some form of rotational Solar power.

At the forefront of this hunt is the revised interest in the 'Stirling' engine, invented and patented by a Scottish cleric in the early 1800s. This engine relies not on bursts of heat that occur at each firing of a cylinder as in an internal combustion engine (normal auto engine), but a *steady* heat applied externally. Tailor-made for Solar heat! This fine engine will turn steady Solar heat into either, a reciprocating motion for pumping, or, rotational motion to perform a multitude of other tasks.

The 'Stirling cycle', as its mode of operation is now termed, works by shuttling a fixed quantity of air between a hot area and a cold area, as a first basic requirement. The second is to ensure that some of the output energy is used to maintain that shuttling.

In a version invented by Dr. Colin West and a team working at the Atomic Research Labs. Harwell, U.K., they have a device that is simpler than the Stirling itself. Theirs has a flexible shuttle (water) and ends up with one version that produces a seesaw action, and another that pumps. The first has only two moving parts and the second none at all. Fig. 130 shows the detail of their so-called 'Fluidyne'.

Much work is also being carried out on the original Stirling engine by Philips in Holland; by several of the automotives in Detroit; and, of particular interest to us, by Eric Farber at the University of Florida. One aspect of Farber's work is aimed at operating the engine directly from Solar; this he does quite

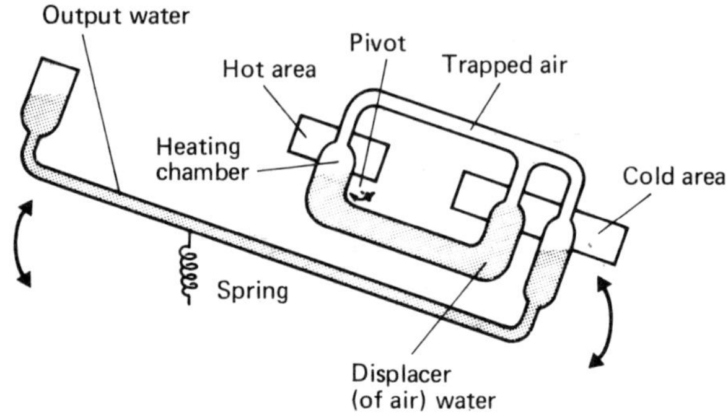

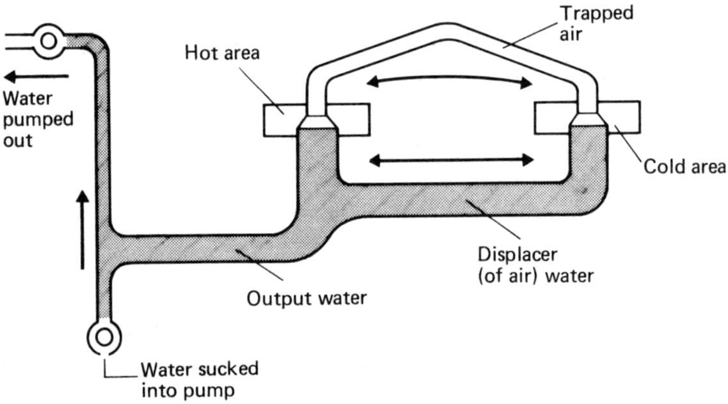

Fig. 130. Two versions of the 'Fluidyne' Stirling cycle engine. *Seesaw pump.* As the hot area warms the trapped air it expands to push the long column of water far enough to the left to overbalance the seesaw and compress the spring. As the heating chamber is now *out* of the hot area, the trapped air cools and the seesaw returns to the 'as drawn' state.

Stationery pump. In this version the displacement crates (used to move the trapped air in and out of the hot area) is not moved back and forth by the seesaw action, but by the pulsations of the output water. The output water pulsates because the trapped air not only moves back and forth but also exbands and contracts. The pulsating does the pumping.

successfully, even to constructing one model from an old lawn-mower engine.

The original Stirling engine (and all others except West's), uses a solid plunger to shuttle the trapped air. Power is produced as the hot air expands and pushes a piston down in the process. A flywheel carries the piston up again, and also moves the plunger to shuttle the air back and forth. Unlike West's version, this engine is quite conventional in appearance.

With the world's second hydrogen conference held in '78, the possibility of a hydrogen based energy concept draws perceptibly nearer. Having previously referred to some of the advantages of hydrogen in an earlier chapter, we show a glimpse of some of the work being carried out to produce hydrogen by Solar. Fig. 131 shows two very simplified sketches of two well established processes. Whilst they are both eminently suited to Solar, they are both very low in efficiency; the present work concentrates on raising that efficiency.

A very recent discovery has used a spinach leaf, and then an 'artificial leaf' as a membrane to produce hydrogen. With input feeds of carbon dioxide, water and Solar Energy, these three inexhaustible ingredients and a simple membrane cell can produce hydrogen.

Another area of constant investigation is storage, and the two most promising new devices are, 'silica gel beds', and, 'two-vessel stores'. The silica gel bed is a new use of these crystals currently under investigation by the James Cook University, Townsville, Australia. Their objective is to make a smaller version of the pebble store. By substituting silica gel granules of 6 mm. diameter in place of 47 mm. diameter pebbles, the same performance can be obtained with a volume *fifteen times* less. As the silica gel is also relatively unaffected by thermal losses, there is also a saving on insulation.

The two-vessel store introduces us to a fresh storage technique as a result of studies carried out by members of the 'Autonomous Housing Study' team under the direction of Prof. Alex Pike at Cambridge University, U.K. The idea relies on the chemical reaction that when acid and water are mixed then

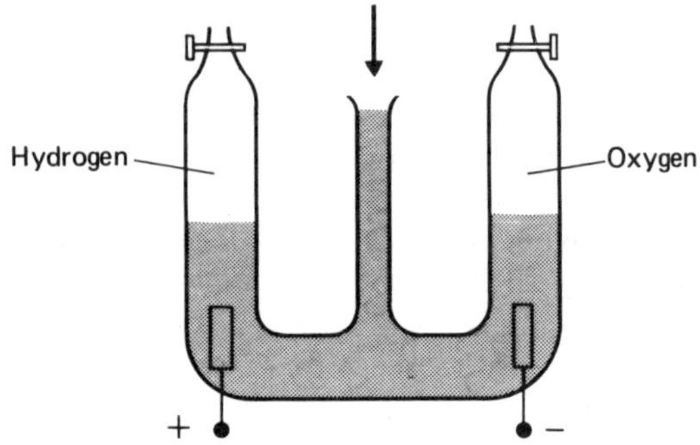

The electrolysis of water is well known – apply the D.C. output from a wind generator and the energy produces hydrogen.

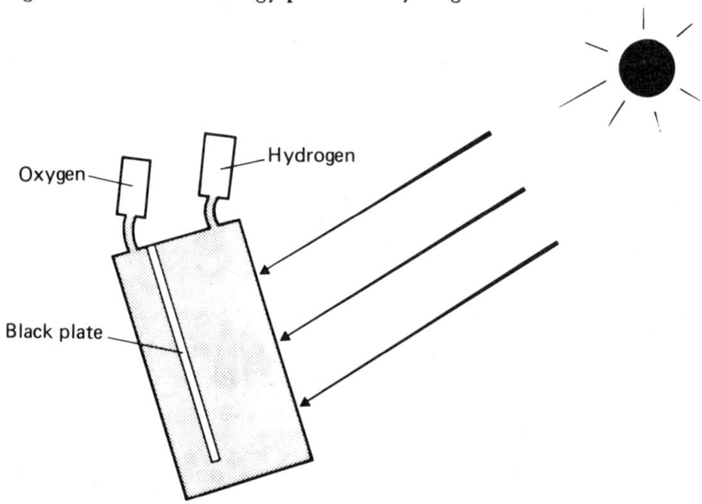

Fig. 131. In the 1937 discovery of Robin Hill, direct sunlight will also break down water to produce hydrogen.

heat is released. The converse also applies, in that if heat is applied to an acid/water mixture, it will separate. Clearly then, if one wishes to store heat, it can be used to drive water and acid into separate vessels where they can remain for years as stored energy. Allow the acid back into the water and the stored heat is released! There are problems but they are being worked on.

A device achieving similar results as does the Beadwall described earlier, is now having the last bugs ironed out of it at the School of Architecture, M.I.T., Cambridge, Mass. They have a window covered with what they call 'cloudy gel' which automatically controls the Solar coming in through the window. When the Sun shines sufficiently to make the room too hot, the gel goes opaque so shutting out the rays. On a cooler winter's day the gel goes clear to allow in the Sun. The main work remaining is to extend the life from the present 3 to 25 years.

Wind generators have at least three new ideas coming along to join them, as shown in Fig. 132. At the top is a 'rocking sail' device that intercepts ground winds in a unique way; the center shows a 'Giromill' version of a vertical axis mill; and the bottom shows a 'floating wing' generator designed to collect the upper winds. This last device cleverly takes advantage of the full wind speeds above 15 meters (50 ft.), without recourse to expensive towers. Recall that getting up from 5 to 20 meters not only finds twice the wind speed, but also 8 times the power!

In the world of collectors nearly all the new devices will be aimed at reducing the cost, although this may appear in the guise of increased efficiency at the same cost. This means that as production increases and the days of handmade collectors pass, the *labour content* of the product will reduce to a minimum. As the only other major production cost is the cost of material, the other move must be towards *cheaper materials.*

While the first change in materials could be from copper to steel or aluminum, the completely corrosion free properties of plastics may well prove an irresistible lure. And although copper and aluminum make excellent devices to heat water we should not forget that they are only intermediaries. The objective is to heat water or air, not metal! At least 2 all-plastic collectors have already appeared, and the combined advantages of: suitability for mass production, cheap raw materials, and long life ensures that many more will follow.

A large Corporation must shortly offer a really low-priced

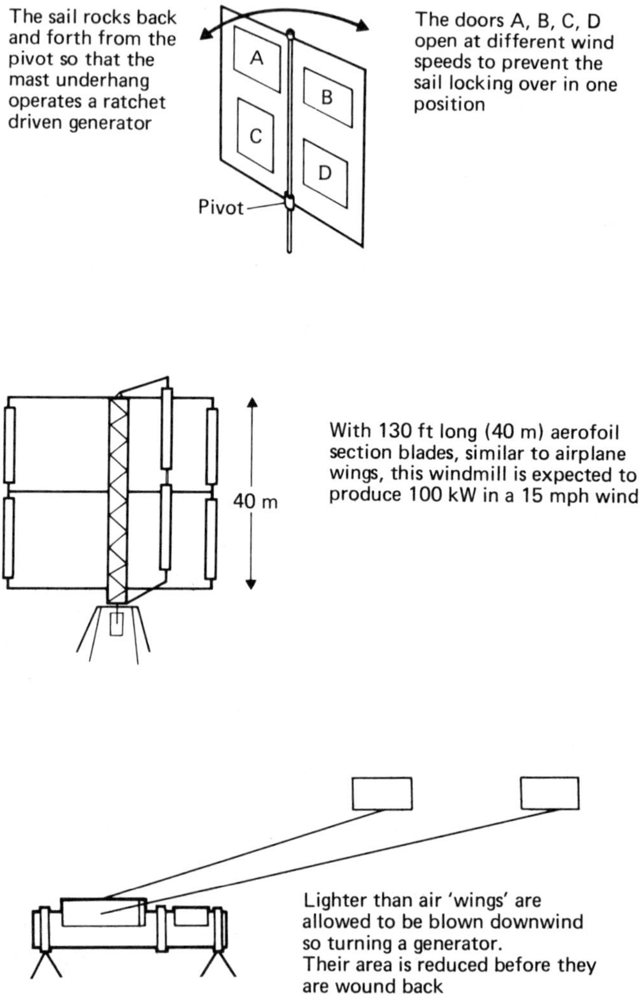

Fig. 132. More wind generation ideas coming along. A variation of the 'floating wing' could be to merely pull the wing down to near ground level before winding it in. This way it goes out on the strong wind and back on the light.

mass-produced product and then the metal collector will go the way of the metal bucket.

So we have traversed full circle back to collectors and costs – strangely enough an area of eternal hope and surprise. Prior to the 1975 World Solar Energy Conference, it was not unusual to hear a favorite exposé of the faint hearted being expounded something like this: "a collector can never get below $ 5.00 a square foot – for a start you *must* have a square foot of gloss, so that's $1.20, plus 2 plates at . . . etc. etc. . . . so it's $5.00 minimum".

It is a fitting tribute to man's unfailing ingenuity that even that basic assumption was shattered at the conference. Replace the glass with a layer of clear fluorescent tubes and the cost is reduced fivefold!

Solar is already doing well – it will do even better.

Directory

The World's Solar industry is now so large and expanding so fast that a full directory of manufacturers, distributors, retailers, installers and advisers would fill this book.

To guide the reader to the latest information in their own country we give the key addresses to write to for information.

As many of these organizations are manned by volunteer labor or run on very limited budgets, we know that a self-addressed and stamped envelope would be much appreciated.

UNITED STATES

'Solar Heating, Cooling & Energy'
Environmental Action of Colorado
1100 14th Street
Denver CO 80202

A 400-page directory about $20

ASHRAE (Circulation Sales)
345 E 47st
New York 10017

A range of publications

Dept. of Energy (formally ERDA)
Technical Information Center
P.O. Box 62
Oak Ridge
Tennessee 37830

Federal data

S.E. Industrial Association
1001 Connecticut Av. N.W.
Washington D.C. 20036

Trade Association

S.E. Digest
P.O. Box 17776
San Diego, CA 92117

Monthly newsletter

I.S.E.S. (American Section)
c/o American Technological Univ.
Killeen, Texas 76541

The International S.E.Society has many Chapters and Regional groups.

CANADA

S.E. Society of Canada
P.O. Box 1353 Winnipeg
Manitoba

Associated to I.S.E.S.

Brace Research Institute
McGill University
Ste. Anne de Bellevue
Quebec

A range of publications.

GREAT BRITAIN

I.S.E.S. (U.K. Section)
19 Albermarle St.
London W.1.

Solar Trade Association
c/o The Building Centre
26 Store St.
London W.C.1.

FRANCE

Centre National de la Recherche Scientific
(C.N.R.S.)
Paris.

GERMANY

German S.E. Society
Postfach 20 0604
D–8000 Munchen 2
Federal Republic of Germany

Associated to I.S.E.S.

ITALY

I.S.E.S. (Italian Section)
Via Crispa 72
Napoli 80121.

HOLLAND

I.S.E.S. (Dutch Section)
c/o Technical Physics Dept.
Utrecht Univ. Utrecht.

SPAIN

C.N.E.E.
INTA–CONIE
Pasco del Pinor Rosales 34
Madrid 8

SWITZERLAND

S.S.E.S
c/o ETHZ Leonhardstrasse 27
CH–8001 Zurich

AUSTRALIA

International Solar Energy Society
P.O. Box 52
Parkville, Victoria
Australia 3052

This is the Headquarters of the Society.
I.S.E.S. Australian Section see New Zealand.

NEW ZEALAND

I.S.E.S. (Australia/New Zealand Section)
CSIRO P.O. Box 26
Highett, Victoria 3190

D.S.I.R.
Lower Hutt
New Zealand

A government research lab.

SOUTH AFRICA

I.S.E.S. (South African Section)
NBRI CSIR
P.O. Box 395
Pretoria 0002.

S.E. Society of South Africa
P.O. Box 309
Estcourt 3310
Natal.

National Building Res. Inst.
Pretoria

Note. The International Society publish 3 items. The magazine *Sunworld,* the monthly journal *Solar Energy* and the quarterly newsletter *ISES News.*

Index